A B C
of Hatchery Management

A B C of Hatchery Management

Professor M. Tufail Banday Division of LPM Faculty of Veterinary Sciences & Animal Husbandry Shere Kashmir University of Agricultural Sciences & Technology of Kashmir, Srinagar (India)
&
Dr. Mukesh Bhakat Senior Scientist National Dairy Research Institute Karnal Haryana (India)

PARTRIDGE
A Penguin Random House Company

ISBN: Softcover 978-1-4828-1653-2
Ebook 978-1-4828-1652-5

To order additional copies of this book, contact
Partridge India
000 800 10062 62
www.partridgepublishing.com/india
orders.india@partridgepublishing.com

Contents

List of Tables

Dedicated

to our

Parents

Preface

Incubation of eggs has been a very specialized task. The avian egg contains a minute reproduction cell quite comparable with that found in mammals. The operation of a hatchery involves the production of the largest number of quality chicks possible from the hatching eggs received at the hatchery. The chicks produced must be produced economically. Ever increasing competition and integration within the industry have made hatchery operation a business of small unit margins and the managers must be ever cognizant of the little things that produce profits.

This book is intended to serve as a guide for use in daily hatchery operations. It contains practical procedures needed for successful incubation of chicken eggs from arrival and quality control upto the placement of day old chicks on the farm.

The book gives practical information on all aspects of hatchery management. it is a unique book because it not only gives theoretical information about incubation of eggs on large scale basis but also provides practical approach in the form of trouble shooting charts on the basis of gross observation of discarded eggs and its diagnosis. There is a section which deals with the important diseases relevant to hatchery borne infections. The trouble shooting chart is unique and would be helpful to provide practical guidelines for field diagnosis and faults in hatchability not only at the flock level but also during incubation.

There is little attempt at capturing all relevant information in a single text. The book is written with the commercial industry in mind because we are aware of this difficult task facing breeders, farmers, managers and technicians of hatcheries in realizing the genetic potential of today's breeding stock. One of its purpose is to provide indepth knowledge about the hatchery management.

It would also fulfill the requirements of students of Veterinary Colleges to cater their needs as per the new curriculum prescribed by Veterinary Council of Indian at undergraduate level and of researchers during their postgraduate studies.

Use this book as a reference, but don't forget to look at the birds

Authors

Foreword

This book '*ABC of Hatchery Management*' written by two Poultry experts Dr. M.T. Banday (Professor & Head), F.V.Sc. & A.H., Sher-e-Kashmir University Of Agricultural Sciences & Technology Of Kashmir and Dr. Mukesh Bhakat (Senior Scientist), National Dairy Research Institute, Karnal, Haryana contains practical procedures needed for successful incubation of chicken eggs from the arrival upto the placement of day old chicks on the farm. This book is designed to provide detailed information regarding embryonic development, hatchery borne diseases and hatchery sanitation. A special feature about the book is the incorporation of photographs regarding the equipments needed for the establishment of a hatchery and daily changes depicted in the development of embryo during incubation has also been provided through a series of photographs. Most of the material it contains is also relevant to the prevailing condition of our country. Since there has been a long felt need for a book that would serve in the long run to the needs of veterinarians, hatchery managers, students of B.V.Sc. & A.H. programme and the new entrepreneurs who want to enter in the hatchery business.

Prof. G.
Devegowda
Date: June 19th 2013

Dr. G. Devegowda B.V.Sc.,M.V.Sc, Ph.D (USA)
Advisor on Animal Biotechnology; Vice President, World Poultry Science Association (IB)
Former Professor and Head, Division of Animal Sciences, University of Agricultural Sciences, Bangalore, India.

743, 7th A cross Sector ' A', Yelahanka Newtown, Bangalore 560064, India.
Tel : + 91- 80-28565282; Mob : +91-9845324750; E-mail: *devegowdag@gmail.com*

1

Introduction

Artificial incubation of poultry eggs is an ancient practice. Aristotle writing in the year 400 B.C. mentions about the Chinese who developed artificial incubation at least as early as 246 B.C. These early incubation methods were often practiced on a large scale, a single location perhaps having capacity of 36,000 eggs. The early Egyptian incubators of some 3,000 old years ago were a series of mud brick egg ovens type rooms built on each side of a central passage way all within a large mud brick building. Thousands of eggs were placed in heaps on the floor of each incubator room. In the upper chamber of the room, there were shelves for low burning fires of straw, camel dung or charcoal to provide radiant heat to the eggs below. The entrance to each incubator room from the passway was though a small manhole.

Temperature control was achieved via the strength of the fires; jute covers over the manholes and regular openings of vents in the roof of the ovens and pass away. Humidity was controlled by spreading damp jute over the eggs, the roof vents also allowed smoke and fumes from the fires to escape and provide some light. The piles of eggs were rearranged and the eggs turned twice a day. The middle pass way also served as a warm brooding area for the chicks after hatching.

The amazing thing about all this was that the temperature, humidity and ventilation were checked and controlled without using measuring devices or gauges nor were there any thermometers. They achieved all this by having the hatchery manager and workers actually living inside the hatchery building. It is recorded that they tested the temperature of the eggs by holding the eggs against their eye lid, the most sensitive part of the body for judging temperature.

The Egyptian hatchery methods were jealously guarded as trade secrets and handed down from generation to generation within certain families in a monopoly situation. Local farmers brought their fertile eggs to the hatchery. The hatchery owner was required by law to return two chicks of every three eggs received, the surplus chicks providing his remuneration.

By the mid 1660's European had bought Egyptian experts to Europe to build and operate an Egyptian type hatchery but were not successful.

After the failure of Egyptian incubators in Europe, the pressure was on to develop a more sophisticated mechanical incubator. The French Scientist, de Beaumur published in 1750, 'The art of hatching and bringing up domestic fowls of all kinds, at any time of the year either by means of hotbeds or that of common fire.'

Over the next 100 years more experimental incubators were produced, some using hot water, some heated by charcoal others by steam. Self regulating oil or kerosene lamps with a damper were also used to heat water and hot air incubators in the 2nd half of the 19th Century. It was the advent of thermostats to regulate temperature accurately which allowed the development of modern incubators.

The two were the development of the electric forced draught egg incubator in the early 1930's. The larger forced draught incubators revolutionized the production of day old chickens not only in the quality of the chicken but also in hatching percentage of eggs set.

The progress in the poultry industry was extremely rapid throughout the world from 1960 onwards. Two outstanding changes occurred. One was the increased size of operations such as the number of birds that could be reared in a single shed and large number of chicks hatched in incubators. The other great advance was technology. This was necessary of the new economies of large-scale production.

Today's automated hatcheries produce upto 1.5 million chicks per week in four equal hatches. To keep pace with demand, production will have to scale-up and in some super hatcheries, about six million day old chicks are hatched every week in routine hatches of a million chicks a day.

2

Breeder Flock Management

Keeping birds for the production of hatching eggs involves practices necessary for the production of table eggs plus practices needed that will produce eggs that are fertile and hatch well.

1.1. BROODING MANAGEMENT:

Production of a good pullet is one of the most important requisites of good management for how well the pullet is grown will greatly determine how well she will produce in the laying house and similarly well managed growth in a male will exert an influence on his behaviour in the breeding pen.

1.2. GROWER MANAGEMENT:

The growing period follows the brooding period and concludes with sexual maturity. Probably no other age of chicken commands more attention more than this period. How well a bird is grown will greatly determine how well it does in the laying or breeding house. Poor quality pullets at maturity will always perform below the standards. Egg production will be poor, egg quality and inferior size of eggs.

1.3. LAYER MANAGEMENT:

It is now estimated that about 75% of all the commercial layers in the world are kept in cages but still a small percentage of commercial layers are kept on a floor of litter, slats or wire. The following points should be considered for increasing the number and the quality of eggs on the floor.

1. The hours of natural plus artificial light per day must be increased when the birds reach sexual maturity.
2. Proper nest management will be a great aid in producing quality eggs and with less breakage.
3. To prevent pullets within the nests overnight, ensure that the birds should be removed from the nests during night hours.
4. Laying nests should be provided in the layer house about one week before the hens start laying eggs.
5. Provide ample clean and dry nesting material in the laying nests.
6. Adequate number of nests should be provided in the house.

FACTORS INFLUENCING THE QUALITY OF EGGS:

It has been shown that the quality of egg is related by a number of factors which are as follows:

1. **Genetics:** Some strains of birds have the ability to produce eggs with better shells.
2. **Position of egg within a clutch:** The first eggs of a clutch have better shell quality than laid later in the clutch.
3. **Length of lay:** The longer the period of egg production the poorer the shell quality becomes.
4. **Temperature:** The higher the environmental temperature, the poorer the quality of egg shells.
5. **Disease:** Certain respiratory diseases such as Infectious Bronchitis and New Castle disease decrease the egg quality.
6. **Humidity:** It has been observed that the moisture level of the air and not the temperature of the air is the critical factor affecting the egg shell quality.

HOW TO REDUCE THE EGG BREAKAGE:

The following steps should be adopted in order to avoid the incidence of cracked eggs.

1. Some strains of egg type layers produce a higher percentage of cracked eggs. Be careful in choosing the right strain of birds.
2. Handle eggs more carefully at the end of laying period.
3. Provide a cushion bumper at the end of the egg collection area of cages.
4. Provide a proper cage space per bird as crowding increases the incidence of cracked eggs.
5. Collect eggs more frequently.
6. Collect eggs more often during the summer months and from older flock.
7. Collect eggs on egg filler flats only.
8. Provide a balanced ration with adequate amount of calcium.
9. Prevent cannibalism. It often causes birds to pick at freshly laid eggs.
10. Reduce bird fright. The jumping of the birds in the cages can create more cracked eggs.

1.4. BREEDER MANAGEMENT:

Keeping birds for the production of hatching eggs involves practices necessary for the production of eggs that are fertile and hatch well. The breeder flock is generally raised in one of the following system as:

- All litter floor.
- Slat and litter floor.
- Cage system.

A poultry farmer has the option of two programs in managing the male and female birds as:

1. **Sex intermingled:** Here both male and female birds are reared together.
2. **Sexes raised separately:** Here the male and female birds are reared separately to reduce fighting in the males.

TOE TRIMMING AND COMB TRIMMING:

In order to prevent the injury to the backs of females during mating, the toes of day old meat type cockrels should be trimmed at the hatchery.

Adult males in breeding pens do a lot of comb pecking to set up their social order. To reduce the comb injury, the combs of male chicks are usually trimmed at the hatchery.

MANAGING THE BODY WEIGHT OF THE BREEDER PARENTS:

The body weight of the meat-type breeders is especially important as they grow very fast and will try to put on more weight thereby affecting their potential to produce maximum number of eggs during their laying cycle. Therefore, controlling the growth of breeder females at the time of sexual maturity should be managed.

Nests: Provide one nest for every four pullets in the breeding house. Nests for meat type breeders should be slightly larger than those for egg type breeders.

Light: An adequate programme of lighting is requisite for maximum hatching egg production.

Prevent floor eggs: Floor eggs are usually dirty and difficult to clean and sanitize and a probable cause of "blowups" in the incubator. A high percentage of them are broken, resulting in a direct loss. There must be a low incidence of floor eggs if breeding operation is to be practical. The number of cracked eggs should never be over 2% for young layers, 3% for old.

Ratio of males to females: Too many males in the breeding pen reduce fertility as do too few. The correct ratio of males to females depends on the type and size of birds involved and is usually recommended as mentioned below:

TABLE 2.1: MALE TO FEMALE RATIO IN BREEDER FLOCK

Type of bird	Male: Female ratio
Mini-leghorn strain	1:10
Medium leghorn strain	1:9
Standard leghorn strain	1:8
Standard meat type strain	1:8 or 9

(Source: Combined management guides of Hubbard, Hybro, Ross)

MANAGEMENT OF MALES FOR FERTILITY:

It is very important to grow a male of high quality as it is to grow a female of similar quality. Too often the male is neglected. Excessive body weight at maturity must be avoided. Any inferior bird must be removed.

EXERCISE THE MALES:

Cockrels should be induced to exercise to prevent their legs from deteriorating.

TIMID MALE:

Males set up a social order as do females. The more timid males must be adequately provided for. Be sure they are getting enough feed to maintain their recommended body weight.

HATCHING EGG PRODUCTION:

The reason for keeping breeding birds is to produce an abundant number of hatching eggs that will produce a high percentage of quality chicks. A good egg usually means a good chick.

HATCHING EGG SIZE:

Hatcheries usually set up minimum weights for hatching eggs that they will incubate. The weight may be different for various lines and breeds. In some cases a lower weight is allowed during the first few weeks of egg production than is allowed later.

WHEN TO START SAVING HATCHING EGGS?

Generally the chicks hatched from the eggs laid by a pullet during her first 2 weeks of egg production do not grow well. But after that, eggs may be used for incubation as soon as they are large enough. The minimum size is determined by the size of the bird laying the eggs.

Eggs produced at the start of lay are smaller than those laid later. Because the same birds are laying them, the first hatching eggs have a genetic potential identical with those produced later. Thus from a genetic standpoint, smaller hatching eggs may be used during first 12 weeks of total egg production than may be used later.

SHELL QUALITY:

Shell quality is associated with hatchability. The longer a bird produces eggs, the poorer the shell quality. Season of the year, strain, temperature, diet and various other factors also affect the texture. Although the quality of the egg shell remains quite acceptable for a 12 month laying period with egg type breeders, shell quality of egg produced by meat type birds begins to deteriorate rapidly after 8 or 9 months of lay.

INTERIOR EGG QUALITY:

Although several of the factors causing fluctuations in the interior quality of eggs also produce variations in hatchability, the important factor is a condition known as tremulous air cells. Some eggs are laid with loose air cells, many more loose air cells are produced during the course of handling the hatching eggs before they are placed in the incubator. Handle eggs carefully and prevent as much as jarring as possible to prevent the loosening of the air cells. Such eggs hatch poorly; many do not hatch at all.

3

Maintaining Hatching Egg Quality—Selection, Care & Storage

One of the miracles of nature is transformation of egg into the chick. In a brief three weeks of incubation, a fully developed chick grows from a single cell and emerges from an apparently lifeless egg. The points to be considered for selection of hatching eggs are given here under:

3.1. SELECTION OF HATCHING EGGS:

Before setting eggs in an incubator, one must obtain or produce quality fertile eggs from a well managed, healthy flock which are fed properly balanced diets. The hatching eggs can be selected as follows:

1. Select eggs from breeders that are:-

   a. Well developed, mature and healthy.
   b. Compatible with their males and able to produce a high percentage of fertile eggs.
   c. Reared under stress free environment.
   d. Fed a balanced breeder diet.

2. Quality of eggs: Eggs are judged for quality by:-

   a. External quality.
   b. Internal quality.

a. EXTERNAL QUALITY:

I. **EGG SIZE:** Select eggs for hatching that are normal in size. For optimum results of hatchability the optimum size of egg should be 56.7 gm. The extra large or small eggs are often infertile. Therefore, incubation of these eggs should be avoided. Large eggs hatch poorly and small eggs produce small chicks. Chicks also lose weight rapidly after hatching because of dehydration so the day old chick weight varies greatly depending on the weight of egg.

II. **EGG SHAPE:** Shape of the egg also plays a vital role in the proper incubation of hatching eggs. Very long eggs as well as very round eggs should be discarded as such eggs do not fit properly into the setter trays, thereby rendering them unfit for incubation. Cracked shells should also be rejected as cracks allow easy entry of microbes in the eggs leading to rotting of such eggs. Many eggs have shell imperfections such as ridges, pointed ends etc. and do not hatch satisfactorily. Some of these imperfections are inherited; therefore, such eggs should not be placed under incubation in order to reduce their incidence in the hatched chicks.

Abnormal shape of the eggs affects the hatchability drastically (Table—3.1)

TABLE 3.1: HATCHABILITY OF ABNORMAL EGGS

Abnormality	% Hatchability
Normal	74
Ridged	65
Round	63
Small	62
Pimpled	19
Wrinkled	13

Source: Mauldin, 1997

III. **SHELL QUALITY:** The quality of the shell is related to hatchability. Eggs possessing strong shells hatch better than eggs with thin shells. The kind of shell depends upon breeding, nutrition and season some strains of chicken produce eggs with thick, strong shells whereas others lay eggs with thin weak shells. The amount of calcium and vitamin D in the ratio affects the shell. Also eggs produced in hot seasons have thinner shells than those produced when the weather is more moderate. For best hatching results egg shells should be between 0.33 to 0.35 mm in thickness. Shell quality depends on the genetics of chicken, nutrition of the breeder flock and the ambient temperature at which the flock is reared. Diets low in calcium and vitamin D feed at 27°C-32°C are more conducive to the production of eggs with inferior shells. Furthermore, the longer a hen remains in egg production, more shell quality will be deteriorated.

IV. **SHELL COLOUR:** The density of the pigment in brown-shelled eggs is often correlated with hatchability. However, hatchability is a genetic factor strains of chicken may be developed that produce high or low hatchability irrespective of egg shell colour. Thus strains of chicken laying eggs with light brown shell would not necessarily involve poor hatchability.

V. **SPECIFIC GRAVITY:** A positive relationship exists between fertility of eggs and their specific gravity.

VI. **SHELL TEXTURE:** Eggs set for incubation must have smooth, thick and uniform texture.

VII. **CLEANLINESS OF EGG SHELL:** Slightly soiled eggs can be cleaned by cleaning the soiled areas with sand paper. However, very dirty eggs should be rejected as these eggs are more prone to microbial proliferation, thus hatchability is reduced in such eggs.

b. INTERNAL EGG QUALITY:

The internal quality of hatching eggs can be determined with the help of candlers. There is evidence that eggs which show quality when candled before incubation for the following parameters. The criteria involved in candling are:

I. Shell quality
II. Condition of air cell
III. Yolk quality
IV. Haugh unit values
V. Other defects

I. **SHELL QUALITY:** While candling, an approximate assessment of the size and shape of the egg is possible. Very small and very large eggs as well as very long and very round eggs can be identified. Besides, cleanliness of the shell can be visualized. Amount of light passing through the eggs gives a direct indication of the thickness (porosity) of the shell. Cracks in the shell which are not visible to the naked eye can be identified while candling. Non-uniform distribution of shell material can also be detected. Very thin shelled or cracked eggs lose their weight faster and hence their quality deteriorates quickly and such eggs are unfit for incubation. Any defect in the shell makes the egg unsuitable for incubation. The defects in the shell identified on candling are:-

1. Thin shelled eggs may be due to calcium deficiency, increased temperature, respiratory tract diseases and physiological disturbances.
2. Mottled shell—Shell appears patchy.
3. Glossy eggs—very thin shelled and transparent eggs.

II. **CONDITION OF THE AIR CELL:** As the egg becomes older there will be increased loss of moisture and therefore, there is an increase in air cell especially when the egg shell is thinner, temperature and air movements are higher and humidity is low. Air cells in eggs to be incubated must be at the broad end firm and immobile. Any defect in the air cell precludes the egg from being incubated. Moving (tremulous) bubbly (many bubbles indicates rough handling of eggs) and displaced air cells and such eggs do not hatch well.

The defects identified in air cells on candling are:-

1. Tremulous air cells.
2. Bubbly air cells.

III. **YOLK QUALITY:** Yolk quality is identified by colour, visibility and movement of yolk. In a fresh egg yolk is held at the centre of the thick albumen. As the egg becomes older, liquefaction of the thick albumen occurs and results in movement of the yolk to the surface. In very old eggs, yolk is very distinct, flatter and sluggish in movement. The defects identified in yolk are:-

1. Addled: vitelline membrane is broken.
2. Rots: Coloured/ colourless.

IV. **HAUGH UNIT VALUE:** The higher the haugh unit reading for albumen quality, the better the hatchability of the eggs. For higher hatchability, the haugh unit value of eggs should be more than 80 in the fresh eggs.

V. **OTHER DEFECTS:**

a. BLOOD SPOTS: During the beginning of the egg production, there might be some bleeding during ovulation and the same will be attached to the yolk. It is likely to be inherited and hence such eggs are not preferred for incubation, although blood spotted eggs may hatch equally well.

b. MEAT SPOTS: When the haemoglobin in the blood spot is reduced, it appears brown in colour and referred to as meat spot. Sometimes epithelial tissue from the oviduct may slough and appears in the albumen which is referred to as meat spot. The eggs with meat spots are not preferred for incubation although they also might hatch equally well.

c. DOUBLE-YOLKED EGGS: Double yolked eggs result due to multiple ovulating followed by the infundibulum engulfing of both the yolks. The eggs are unfit for incubation, since there is insufficient space and nutrition and also because the size of the egg is too big making it unsuitable for setting in the incubation trays.

d. HEAT SPOT: In case of fertile eggs held at or above a temperature of 68°F the embryos grow but soon die bearing partially developed spots. Such eggs are not suitable for incubation.

e. BLOODY EGG: Blood in albumen may be due to internal injury in the oviduct and subsequent bleeding. Such eggs show red-tinged albumen upon candling and are unfit for incubation.

f. EGG IN EGG: A fully formed egg due to disturbance or some other unknown reasons sometimes moves backwards up to infundibulum by antiperistaltic movements of the oviduct. One or more ovulation and other processes follow resulting in egg in egg. Such eggs are unfit for incubation mostly because of their abnormal size, shape and difficulty in the respiration of the embryos.

g. **MIS-SHAPEN EGGS:** Mis-shapen eggs results due to several factors including respiratory diseases. Such eggs are unfit for incubation mostly because of their abnormal size, shape and difficulty in the setting of eggs in setters.

3.2. CARE OF HATCHING EGGS PRIOR TO INCUBATION:

Once eggs are laid they must be held for a day or more to fit the setting schedule of the incubator. Sometimes the hatching eggs may not be set for 1 or 2 weeks after they are laid. The conditions under which eggs are held have a great bearing on the hatchability of the eggs. The following practices for care of hatching eggs are recommended:-

1. **GATHERING EGGS FREQUENTLY:** Generally hatching eggs are gathered more frequently than eggs intended for table use. When the environmental temperature is normal, 3-4 gatherings a day will suffice, however when the temperatures are extremely hot or cold, hatching eggs should be gathered every hour. Frequently gathering reduces the contamination of eggs from contact with nesting material and faeces and prevents chilling in winter and overheating in summer.
2. **GATHERING CLEAN EGGS:** It is best to collect only clean eggs for hatching purpose. In case of soiled eggs use a non-recycle washer keeping the washing water temperature lower than the egg temperature. Cool the eggs quickly after washing.
3. **SANITIZE THE EGGS:** For effective sanitization, eggs should be treated within 1-2 hours after they are gathered. Several decontaminants may be used as under:-

    a. **FORMALDEHYDE GAS:** Triple strength (6 gm of potassium permanganate with 120 ml of formaline) for each cubic foot of space in the room is usually recommended with the fumigation time in an airtight chamber not exceeding 20 minutes.

b. **QUATERNARY AMMONIA:** It is sprayed on eggs in a lukewarm water solution containing 200 ppm chemical.
c. **CHLORINE DIOXIDE:** It may be used as a spray or as a dip on the eggs soon after gathering.
d. **OZONE (O3):** When generated at 100 ppm, ozone (O3) is an effective sanitizer.

Following laying, the number of bacteria on the shell increases tremendously, therefore, any sanitizer should be administered as soon as possible after the eggs are laid.

4. **EGG HOLDING TEMPERATURE:** Although the optimum temperature for embryonic development in the forced draft incubator is 99.5 ^{O}F (37.5 ^{O}C). Generally at a temperature of 75 ^{O}F (24 ^{O}C) above which embryonic growth commences and below which it ceases. This temperature is called threshold temperature. After hatching eggs are laid they should be cooled to a temperature well below the threshold of the embryonic development when hatching eggs are held at a temperature of 65 ^{O}F, embryonic development is fully arrested. Eggs held for less than 5 days show little perceptible reduction in their hatchability or in the quality of the chicks hatched from them. When the period of holding is longer than 4 days, hatchability will drop considerably with each additional day.
5. **HOLDING MINIMUM TIME:** Hatching eggs should be held for as short a period of time as possible. Hatchability decreases as the time of holding is increased. A thumb rule is for every day eggs are held or stored after 4 days, hatching time is delayed 30 minutes and hatchability is reduced by 4%. At the most, the eggs should not be held longer than 10 days.
6. **HOLDING TEMPERATURE:** When the egg is laid and its temperature drops the development of embryo ceases. As soon as possible, hatching eggs should be cooled to a temperature of 65°F. When it is necessary to hold eggs longer than 7 days, it is recommended that they be warmed to 100°F for 1 to 5 hours early in the holding period.
7. **HUMIDITY:** Moisture held within the contents of the egg is continuously lost by evaporation through the shell. The rate of this process is governed in part by the relative humidity of

the air surrounding the egg. When the relative humidity is low, evaporation of the egg contents is more rapid, when it is high, evaporation is less rapid. High humidity in a hatchery tends to prevent evaporation and an enlargement of air cell and improvement in hatchability. A relative humidity of 75 to 80% is recommended.

8. **POSITION:** Hatching eggs are best held with the small ends up from the standpoint of increasing hatch. However, most hatchery operations do not consider this holding position practical because eggs are incubated with the large ends up, hence eggs that are stored with the small ends up must be inverted.
9. **TURNING:** It is necessary to hold eggs for hatching for more than 7 days, they should be turned. This prevents the yolk from sticking to the shell. Turning can be done by tipping the egg cases sharply. It is recommended that the cases be turned in this manner twice daily.

4

Factors Affecting Hatchability

Process of hatching a chick is influenced by a number of factors which may be categorized into:-

1.1. Pre-incubation factors.
1.2. Incubation factors.

Hatchability may be measured by two formulae:-

1. The number of chicks hatched as a percentage of all eggs set.
2. The number of chicks hatched as a percentage of the fertile eggs set.

4.1. PRE-INCUBATION FACTORS:

Several pre-incubation factors have been shown to affect the hatchability of egg which has been discussed here under:-

a. **NUTRITION OF BREEDER HEN:** A balanced diet feeding to dam is important for proper hatchability of eggs. The egg must contain all the nutrients needed by the embryo when it is laid by the hen. There is no further contact with the mother once the egg is laid. Therefore, breeder hens must be fed rations that will supply adequate quantities of the nutrients needed for the development of embryo. Since it is difficult to affect the protein, fat and carbohydrate content of an egg by dietary means, the nutrients most susceptible to diet changes are the vitamins and trace elements.

Nutrient deficiencies can reduce hatchability and often malformed embryos grow as a result of the nutrient deficiency. It is rather difficult, however to identify the nutrient deficiency responsible for poor hatchability by examination of the embryo. The time of embryo mortality and the deficiency observed often depends on the degree of deficiency of the nutrient involved.

b. **GENETIC FACTORS:** Genetic factors play a definite role in the hatchability of eggs. Inbreeding has been shown to lower hatchability. Some inbred lines are affected to a great extent than others by inbreeding. Certain lethal genes may cause death of the developing chick before the end of incubation. Many of these lethal genes are associated with specific morphological features recognizable in the embryo. Some of the malformed embryos may hatch but they may not be able to survive. These mutations when present may affect hatchability of eggs from a breeder flock. Since several factors, including egg structure and genetic constitution of the embryo affect hatchability.

c. **DISEASES:** Diseases caused by Salmonella organisms such as pullorum disease are the major group of bacterial infections that influence hatchability. Salmonella organisms may be passed from infected dams into eggs. The infected eggs do not hatch as well as non-infected ones. Although other disease organisms may not pass into the egg and affect the embryo directly, they may influence the characteristics of the egg and thus indirectly affect its hatchability. New castle disease and infectious Bronchitis may affect egg shape and shell porosity. Eggs from hens affected by these diseases frequently do not hatch well because the eggs lose excessive amounts of moisture during incubation. Therefore, hatching eggs from healthy flocks are more likely to produce the best hatch.

d. **AGE OF BREEDER FLOCK:** As the age of breeder flock advances, hatchability drops. Their eggs become much larger and are held in the oviduct for longer periods, thereby increasing the length of the preoviposital incubation period leading to more advanced state of development for embryos at the time the egg is laid a period not conducive to holding prior to incubation. Similarly the egg shell of older hens is always thinner especially in hot weather. These large eggs on incubation show a higher incidence of embryonic deaths at the time they are placed in the incubator when embryonic growth is reinitiated. These deaths come so early that they are often not noticed and are usually classified as infertiles. Hatching eggs from older hens should be gathered more than those from younger hens.

e. **SEASON:** Lower hatchability from eggs laid during the periods of extremes in environmental temperatures is common. Continuous days of hot and cold weather are likely to cause a drop in hatchability because hot and cold affect the breeders producing eggs for periods of short duration (1 or 2 days) will not affect. Hot weather during the summer months is a detrimental to good hatchability.

f. **EGG LAYING PATTERN:** Eggs produced initially in the laying cycle do not hatch well for the first two weeks of egg production and the chicks hatched out from such eggs show poor growth. Usually these eggs are held in the hen for a period longer than normal and the pre-incubation is detrimental to hatchability. Therefore, under normal conditions hatching eggs produced during the first two weeks of egg production are not set for hatching. Eggs produced at the end of the laying cycle do not hatch well. Normally, there is a pattern of increased hatchability from first egg set until about 13th week of egg production after which hatchability gradually decreases. Generally eggs from the birds with a high rate of lay, hatch better than those from the birds laying at a medium or low rate.

g. **SHELL QUALITY:** Poor shell quality has probably the greatest effect on the hatchability. Shell quality from young breeder flocks is usually good and its hatchability is high but as the laying continues, the shell quality deteriorates thereby dropping the hatchability.

4.2. INCUBATION FACTORS:

Several factors during incubation have a significant influence on the hatchability of the eggs. These factors have been discussed in detail here under.

a. **TEMPERATURE:** Embryos are poikelothermic during the first 18 days of incubation and hence they are very sensitive to the fluctuations in incubator temperature. Maximum temperature causing little effect on hatchability seems to be 40°C and temperatures above these cause detrimental effects on hatchability depending on the severity and duration. Embryos subjected to heat stress exhibit clubbed, wiry down and unsteady gait. Embryos similar to adults tolerate cold better than hot and during first 19 days of incubation, reducing the temperature to as low as 18.3^{O}C does not seriously affect hatchability. However, since the embryo changes its temperature in time with that of incubator, cooling during the first two weeks is more detrimental than during the next 5 days. In any case cooling lengthens, the incubation period and the effect is cumulative. Besides, cooling during first 19 days also increases the embryo malpositions. Reduction in temperature during the last 3 days is highly detrimental since the birds are changing over to homeothermy and beginning their pulmonary respiration to accelerate the heat production. Development of hypothalamus is also affected due to cooling during the last 3 days.

b. **HUMIDITY:** The relative humidity corresponds to the deviation in readings between dry and wet bulb thermometers. If the humidity is lower than the required 55% during the first 19 days, there will be excessive evaporation of the contents and the chicks will be smaller and drier than normal. Similarly, if the humidity is higher than the required, there will be reduced moisture loss resulting in chicks which are wet and larger than normal. In both cases, the embryo is weakened and hence the chick quality and their survivability will be lowered. Besides, higher humidity results in earlier hatching and vice versa.

Effects of both temperature and humidity in general, and humidity in particular are related to egg weight (size) and shell quality. Smaller eggs have higher relative surface area and therefore, more evaporation of moisture takes place from them. Shells with microscopic cracks or eggs with cracks during setting must be removed from the setter.

Humidity in the hatcher must be increased gradually and optimum humidity may vary between machines. Humidity required during hatching seems to be between 65 to 75 percent and lower humidity will produce chicks smeared with egg or shell stuck down and partial dehydration whereas too much humidity will cause chicks to be smeared with egg contents and unclosed navals.

c. **VENTILATION:**

The oxygen requirement is increased nearly 90 times and carbon dioxide expelled is about 80 times by the end of incubation period. On 18th day of incubation 1000 eggs require 850 lits of oxygen which means 4000 lits of fresh air is needed for the purpose. The air in the incubator should be changed about 8 to 9 times a day or once every 3 hr. However, in forced draft incubators the problem of oxygen availability is considerably rare. Although the oxygen content of the air in a commercial incubator is not often altered, there may be some variation in the hatcher where large amounts of carbon dioxide is given off.

Carbon dioxide is a natural by-product of metabolic process during embryonic development beginning during gastrulation. In fact, CO2 is being released through the shell at the time the egg is laid, the result of the pH of egg contents changing to an alkaline condition. Carbon dioxide concentrates in the air within the setter and hatcher when there is insufficient air exchange to remove it. Young embryos have a lower tolerance level to CO2 than older ones when measured as the concentration in the air within the machines. The tolerance level seems to be linear from first day of incubation through the 21th day. During the first 4 days in the setter, the tolerance level of CO2 is 0.3%. The level above this reduces hatchability with significant reductions at 1.0% and completely lethal at 5.0%. Chicks hatching in the hatcher give off more CO2 than embryos contained in eggs and the tolerance level in the hatcher has been set at 0-75%.

Speed of air flow does not seem to influence hatchability. During the first 13 days of incubation, embryos require heat whereas after 13 days, there will be a necessity of heat dissipation and hence, air flow becomes critical especially after 13 days of incubation.

d. **AIR PRESSURE:** At high altitudes, atmospheric pressure reduces and therefore air expands resulting in reduced oxygen concentration. This results in reduced hatchability and for every 300 m of elevation above sea level, there will be 2.5 percent reduction in weight of air and reduction in hatchability becomes particularly perceivable above an altitude of 1050 m. Hatchability reduces by about 5% for every 300 m elevation above 1050 m. Under such conditions, oxygen may have to be injected into the incubators and it is recommended to maintain an oxygen concentration of 23 to 23.5 percent.

e. **LIGHT:** Light stimulates embryo growth in an unknown way. White light of wave length around 295 milli micron used continuously by fixing the bulbs close to the top of the large end of the egg (bulbs should not move during turning of egg) has been found to advance the hatchability by 20 to 48 hr (depending on the egg size), improve the chick weight by 15% (especially in larger eggs), reduces embryo mortality, improves the body weight of broilers at market age and advances age at sexual maturity in leghorn pullets. However, it does not influence either egg production or male fertility.

f. **POSITION AND TURNING OF EGGS:** The normal position of eggs to be set in the setters is with broad end up to help development of head of the chick in the large end. If set with small end up, they develop head in the small end and fail to break the air cell while hatching. The embryos orient themselves with head upper most near the air cell during the second week of incubation.

Eggs have to be turned back and forth along their long axis. They must not be turned in a circle since such a rotation can result in rupture of allantoic sac. Most eggs are turned to a position of 45° from vertical, then reversed in the opposite direction to a similar position. Rotation through an angle less than 45° is not adequate to avoid embryo mortality due to sticking of yolk to the shell. Although, embryos can come out of the shells kept broad end up in the hatcher, it is not practicable and therefore, eggs are preferably kept in a horizontal position to help the chick to come out of the shell.

4.3 FERTILITY:

The ability of the hens to produce fertile eggs depends on various factors.

1. **Proper management of breeder flock.**
2. **Genetics:** Varieties and strains of chicken differ.

3. **Presence of lethal genes:** Some lethal genes are associated with sex, reducing the hatchability of one sex more than the other.
4. **Physical factors:** Evidently one sex is better able to adjust itself to the environmental conditions of incubation.
5. **Proper sex ratio (Table 4.1):** Whether the eggs are fertile or not it is not possible to determine in an intact egg without incubation, although Japanese claim to have developed an electronic device to identify the sex of the embryo as well. Fertility is inherited but its heritability value is low (0.05). Individual males and females vary in their ability to produce viable embryos. This may depend both on the quantity and quality of semen, mating ratio and mating efficiency etc. Certain mutations are correlated with infertility. Homozygosity of the R gene for rose comb is associated with poor fertility in males but not in females.

TABLE 4.1: SUGGESTED MATING RATIOS

Cockerel	Pullet	Males per 100 females	
		On litter	On slat & litter
Mini Leghorn	Standard Leghorn	8	9
Standard Leghorn	Standard Leghorn	8	9
Medium size	Medium size	9	10
Standard meat type	Mini meat type	9	10
Standard meat type	Standard meat type	10	11

(Source: Combined management guides of hubbard, Hybro, Ross)

5

Egg Formation & Chick Embryo Development

The avian egg consists of a minute reproductive cell comparable with that of mammals. This reproduction cell is surrounded by yolk, albumen, shell membranes, shell and cuticle. The ovary is responsible for the formation of the yolk; the remaining portions of the egg originate in the oviduct. At the time of early embryonic development, two ovaries and two oviducts exist but the right ovary and oviduct atrophies leaving only the left ovary and oviduct at hatching. Prior to egg production the ovary is a quite mass of small follicles containing ova. Some ova are large enough to be visually seen, others require microscopic magnification. Several thousands are present in each female chicken.

5.1. FORMATION OF THE YOLK:

The yolk is not the true reproductive cell but a source of food material from which the minute cells (blastoderm) and its resultant embryo partially sustain their growth. When the pullet reaches sexual maturity, the ovary and the oviduct undergoes many changes. About 11 days before she is destined to lay her first egg, a sequence of hormonal activity takes place. The follicle stimulating hormone (FSH) produced by the anterior pituitary gland causes the ovarian follicles to increase in size. In turn, the active ovary begins to generate the hormones estrogen, progesterone and testosterone. Higher blood plasma levels of estrogen initiate development of medullary bone, stimulate yolk protein and lipid formation by the liver and increase the size of oviduct, enabling it to produce albumen proteins, shell membranes, calcium carbonate for shell and cuticle. The first yolk to matures as major amount of the yolk material produced in the liver and is transported by the blood goes directly to it. A day or two later the second yolk begins to develop and so on until at the time the first egg is laid from 5 to 10 yolks are in the process of growth. About 10 days are required for an individual yolk to mature. Deposits of yolk materials are very slow at first and light in colour. Eventually the ovum reaches a diameter of 6 mm at which time it grows at a greater rate. A greater number of yolks are under development at one time in the broiler breeder hen than in the egg-type hen but the broiler breeder hen does not have the viability to produce as many complete eggs, therefore, she produces fewer. The addition of fat and protein to laying ration has been shown to increase the size of the developing yolk. Ova vary greatly in size not only those produced by an individual chicken but those produced by the various hens in the flock. Size is not associated with the rate of lay but more probably with the length of time required for ova to reach maturity. The yolk material is laid down adjacent to the germinal disc that continues to reach on the surface of the globular yolk mass. Once the egg is laid, the yolk rotates so the germinal disc is uppermost.

5.2. OVULATION:

At maturity the ova are released from the ovary to enter the oviduct by a process known as ovulation. Each ovum is attached to ovary by a narrow stalk containing the artery that supplies the blood to the developing yolk. When an ovum is mature, the hormone progesterone produced by ovary excites the hypothalamus to cause the release of the leuteinizing hormone (LH) from the anterior pituitary which in turn causes the mature follicle to rupture at stigma to release ovum from the ovary. It is not known what sets the hour for the bird's first ovulation, but both the nervous system and hormonal secretions are of primary importance. The second ovulation is regulated by the laying of the first egg and occurs about 15 to 40 minutes after the first egg passes through the vent. Chicken lay eggs on successive days known as clutches after which they will not lay for one or more days. The length of the clutch may vary from 2 days to more than 10 before a day is missed. The time necessary for an egg to transverse the oviduct varies with individuals. Most hens lay successive eggs with time intervals of 23 to 26 hours. If the time is greater than 24 hrs, each successive egg will be later in the day and the ovulation of the yolk for the next egg will also occur later in the day. Hens that produce long clutches lay their first egg of a clutch early in the day, an hour or two after the sun rises or the artificial lights go on. Most ovulations occur during the morning hours. Light either natural or artificial has a stimulating effect on the pituitary gland forcing it to secrete an increased quantity of the FS hormone which in turn activates the ovary. Both duration and intensity of light are important.

5.3. PARTS OF THE OVIDUCT AND THEIR FUNCTIONS:

The oviduct is long tube through which the yolk passes and where the remaining portions of the egg are secreted. Normally it is relatively small in diameter but with the approach of the first ovulation its size and wall thickness expand greatly. The segments of the oviduct and their purpose are summarized in the Table 5.1.

TABLE 5.1: FUNCTION OF DIFFERENT PARTS OF OVIDUCT IN FORMATION OF EGG

Sl.No.	Name of Segment	Length	Functions	Time Spent
1.	Infundibulum	9 cm	After ovulation the yolk is picked up by infundibulum.	15 minutes
2.	Magnum	33 cm	All the four parts of albumen are secreted.	3 hours
3.	Isthmus	10 cm	Inner and outer shell membranes are formed.	75 minutes
4.	Uterus	10-12 cm	Water and salts are added through the shell membranes by osmosis. Egg shell calcification takes place.	18-20 hrs
5.	Vagina	12 cm	Here cuticle is deposited on the shell to fill many of the shell pores.	Few minutes

5.4. SHAPE AND SIZE OF THE EGG:

5.4.1. SHAPE: Although most eggs are ovoid in shape, the exact shape of the egg is generally due to inherited genetic factors. Each hen lays successive eggs of the same shape.

Specifications for a standard egg:-

Weight: 56.7 gms
Volume: 63.0 cm^3
Specific gravity: 1.09
Shape index: 74.0
Surface area: 68.0 cm^2

Some hens continuously lay eggs with shape imperfections. These may be wrinkled, ridged, flat sided, pointed tip etc. Some of the defects may be of genetic origin or probably due to abnormalities of the oviduct.

5.4.2 EGG SIZE:

Eggs from a flock of chicken vary in size (or weight) because of many circumstances. The exact cause of some of the variations is not known. Some of these variations are as follows:

1. Some hens lay eggs that are larger or smaller than those laid by other hens. The difference is mainly due to genetic factors that have an effect on the length of the growth period of the ova.
2. The first egg laid by a hen is smaller than laid later as the egg size gradually increases as the hen continues to lay.
3. The sequence of eggs within a clutch affects egg size. In most instances, the first egg of the clutch is the heaviest.
4. Some feed components will affect egg size e.g., egg size may be increased by increasing the protein content of the feed.
5. Hot weather affects the flock causing a decrease in the egg size.

5.5. DEVELOPMENT OF THE CHICK EMBRYO:

Unlike that of the mammal, the avian embryo develops from the food material stored in the egg rather than from nutrients derived from the blood supply of the mother. Furthermore, most of the embryonic growth takes place outside the body of the mother and development is more rapid than in the case of the mammalian embryo.

5.5.1. FERTILIZATION:

Normally fertilization is a natural process but artificial insemination is also practiced.

a. NATURAL FERTILIZATION:

The cock initiates sexual activity in chicken through a process of courting. Chicken are polygamous but certain cocks and hens mate regularly. During the course of normal mating between a cock and hen the cock ejaculates about 1.5-8.0 billion sperms per ejaculate. First ejaculate averages about 1.0 ml but after several matings they will be reduced to 0.5 ml or less.

A cock may mate from 10-30 times a day depending on the availability of hens and competition from other cocks. As matings continue the volume of semen and the number of sperm cells decrease but an ejaculate will not contain less than 100 million sperms.

Factors affecting fertility:-

1. Movement of spermatozoa in the upper part of oviduct.
2. Time taken by the spermatozoa to reach the area of fertilization.
3. Optimum ambient temperature (19°C) for normal activity of male testes.
4. Age of the sperms produced.

b. ARTIFICIAL INSEMINATION:

It is possible to obtain semen artificially from the male chicken to inseminate females. Generally one male is required for about ten females but with artificial insemination one male can produce enough semen to fertilize 100-150 females on weekly basis. Fertility through artificial insemination is in leghorns than in meat type strains.

Procedure: The soft part of the abdomen below the pelvic bones of the male is massaged to protrude the papillae and the semen is gently squeezed out, beginning closest to the body and collected in a vial. The semen is then transferred to a syringe, diluted with special diluents. About 0.025 to 0.035 ml is inseminated into the oviduct of the hen to a depth of about 2.5-5.0 cm depending upon the size of the bird. Males should be ejaculated about 3 times per week. Semen collected in the morning will have greater volume, greater sperm motility and slightly higher sperm concentration than that collected in the afternoon. Insemination should be done quickly after fresh semen is collected. It will not withstand freezing and fertility will drop to about one-half.

5.5.2. DEVELOPMENT OF EMBRYO:

The development of chicken embryo can be discussed under the following two headings:

a) Preoviposital embryonic development.
b) Postoviposital embryonic development.

a) PREOVIPOSITAL EMBRYONIC DEVELOPMENT:

The first embryological development takes place within the body of the hen at her body temperature between 40.6°C and 41.7 °C. About 4.5% of the total consumed in embryological development takes place in the oviduct. In general the average total incubation process requires about 22 days: 1 day in the hen and 21 days in the incubator.

Preovipositai development is initiated in the infundibulum about 15 minutes after the yolk is ovulated when the sperm cell enters the female egg cell to form the zygote. About 5 hours later, the new zygote enters the isthamus portion of the oviduct and the first cleavage takes place and two daughter cells are formed. About 20 minutes later, the two new cells divide and four cells are produced. By the time the forming egg leaves the isthamus, eight cells are produced to enter the uterus. After 4 hours in the uterus the developing embryo grows to 256 cells.

During the above divisions of the cells the blastodisc is formed. As cellular divisions continue, several layers of cells develop and make up the blastoderm. Soon the cells in the centre of blastoderm become detached from the yolk to form a pouch or a cavity. It is in the centre of this cavity that further embryonic development takes place.

While the developing egg is still within the body of the hen the blastoderm develops into two layers by what is called gatrulation. The upper layer of cells is called the ectoderm and the lower is called as entoderm. Soon a third layer, the mesoderm develops between the ectoderm and entoderm. From these three layers all the organs and the parts of the body develop as follows:-

(I) Ectoderm gives rise to

i. Nervous system,
ii. Parts of eyes,
iii. The feathers, beak, claws and skin.

(II) Endoderm gives rise to respiratory and secretory organs

i. Digestive tract

(III) Mesoderm gives rise to

i. Skeleton,
ii. Muscles,
iii. Blood system,
iv. Reproductive organs
v. Excretory system.

By the time the egg is laid, the developing embryo will be composed of thousand cells.

(b) POST OVIPOSITAL DEVELOPMENT:

In a fertile egg just laid, the embryo should be developed past the gastrula stage and well suited to a cessation of development prior to being placed in the incubator. To arrest all development during the holding period the egg should be maintained at a temperature of 23.9^{O}C, when held for less than 5 days. When held for a longer period the hatching eggs should be kept at 12.8 °C.

5.5.3. DEVELOPMENT OF EXTRA EMBRYONIC MEMBRANES:

Embryo has no anatomical connection with the mother's body; natu4re has endowed it with certain membranes necessary to utilize the food material contained in the egg. These membranes are as follows.

a. **YOLK SAC:** Enveloping the yolk, this membrane secretes an enzyme that changes the yolk contents into a soluble form so that the food material may be absorbed and carried to the developing embryo. The yolk sac and it's remain contents are drawn into the body cavity just prior to hatching to serve as a temporary source of food.
b. **AMNION:** The amniotic sac helps the young embryo during development as it is filled with a transparent fluid in which the embryo floats.
c. **ALLANTOIS:** This membrane serves as a circulatory system and when fully developed completely surrounds the embryo. The allantois is initiated on the third day and is fully developed by the 12th day. It has the following functions:

    I. **Respiratory:** It oxygenates the blood of the embryo and removes the carbon dioxide.
    II. **Excretory:** It removes the excretory material from embryonic kidney and deposits them in the allantoic cavity.
    III. **Digestive:** It aids in the digestion of albumen and in the absorption of calcium from the egg shell.

d. **CHORION:** This membrane fuses the inner shell membrane with the allantois and helps the latter in completing its metabolic functions.

5.6. DAILY CHANGES DURING EMBRYONIC GROWTH:

During the incubation, moisture is lost from the egg through the shell. This drying reduces the size of the egg contents and increases the size of the air cell. After 19 days of incubation the air cells usually occupies one third of the egg. It is deeper on one side than on the other. The development of the chick embryo is a complicated process. The main changes that occur during the process of incubation are given in the Table 5.2 and diagrammatically represented in Fig 5.2.

TABLE 5.2: DAILY CHANGES THAT OCCUR IN EMBRYO DURING INCUBATION

Age in days	**Changes**
1st day	Several embryonic processes are evident during the first 24 hrs of incubation.
4 hrs	Heart and blood vessels start to develop.
12 hrs	Heart starts to beat. Blood circulation begins with the forming of blood vessels of embryo and the yolk sac.
16 hrs	First sign of resemblance to a chick embryo with development of somites from which bones and muscles develop.
18 hrs	Appearance of the alimentary tract.
20 hrs	Appearance of the vertebral column.
21 hrs	Origin of nervous system.
22 hrs	Head begins to form.
24 hrs	Eyes originate.
Appearance development of germinal disc 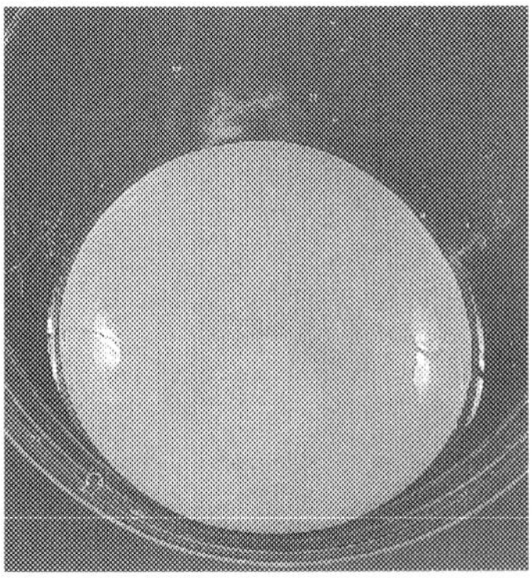	

2nd day	Beginning of tissue development is visible 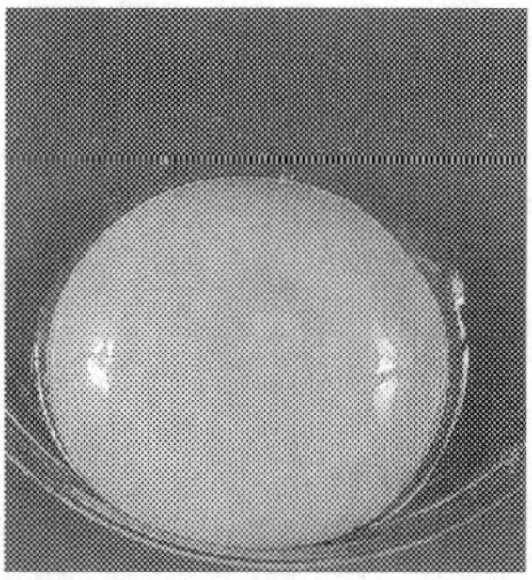
25 hrs	Beginning of the formation of the ear.
3rd day	Blood vessels are very visible 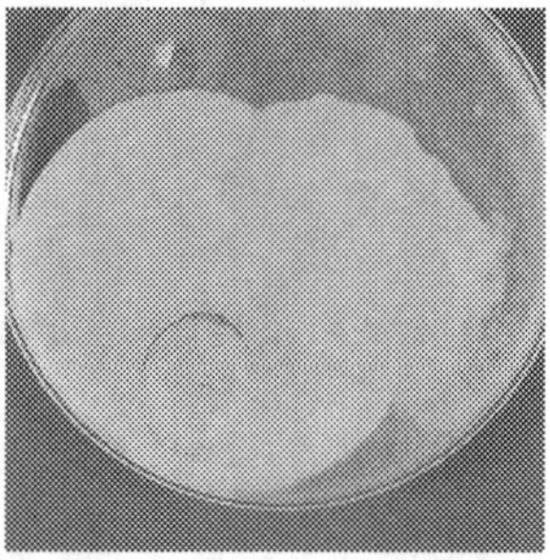
60 hrs	The nose is initiated.
62 hrs	The legs begin their development.
64 hrs	Beginning of the formation of the wings. The embryo begins to rotate and it lies on its left side. The circulatory system rapidly increases.
4th day	The tongue begins to form and now all body organs are present. The vascular system is evident to the naked eye. Eye is pigmented 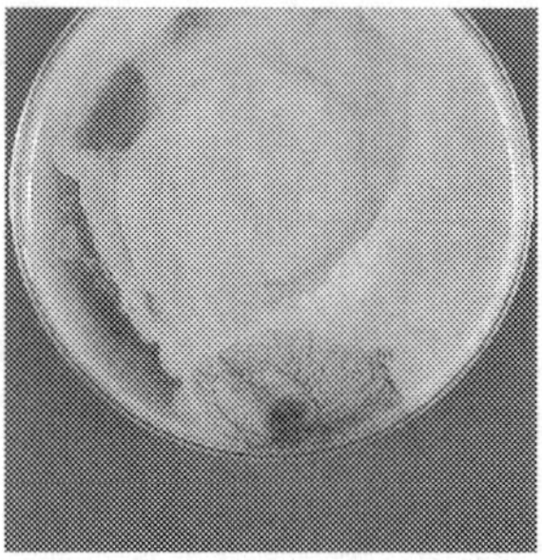

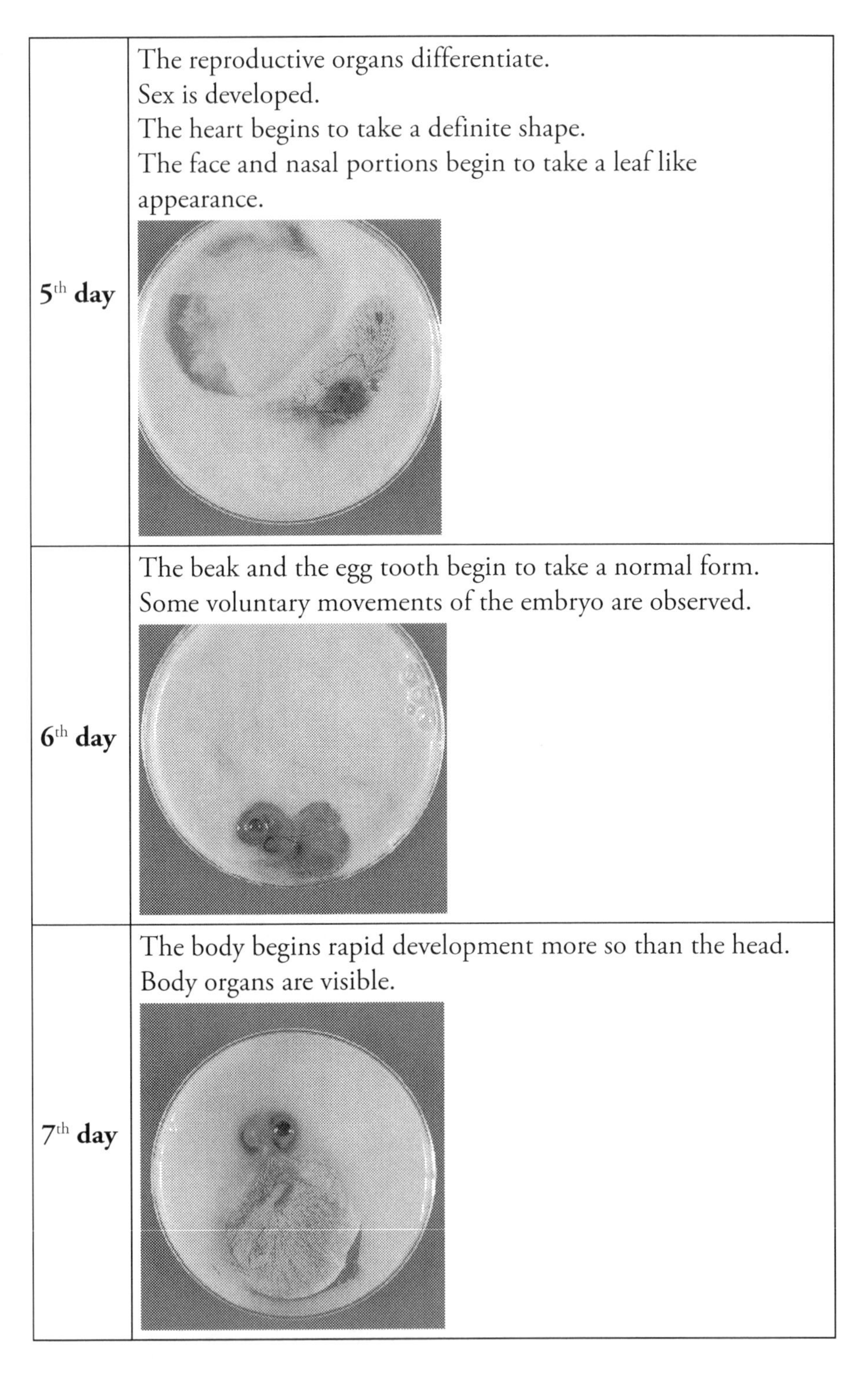

5th day	The reproductive organs differentiate. Sex is developed. The heart begins to take a definite shape. The face and nasal portions begin to take a leaf like appearance.
6th day	The beak and the egg tooth begin to take a normal form. Some voluntary movements of the embryo are observed.
7th day	The body begins rapid development more so than the head. Body organs are visible.

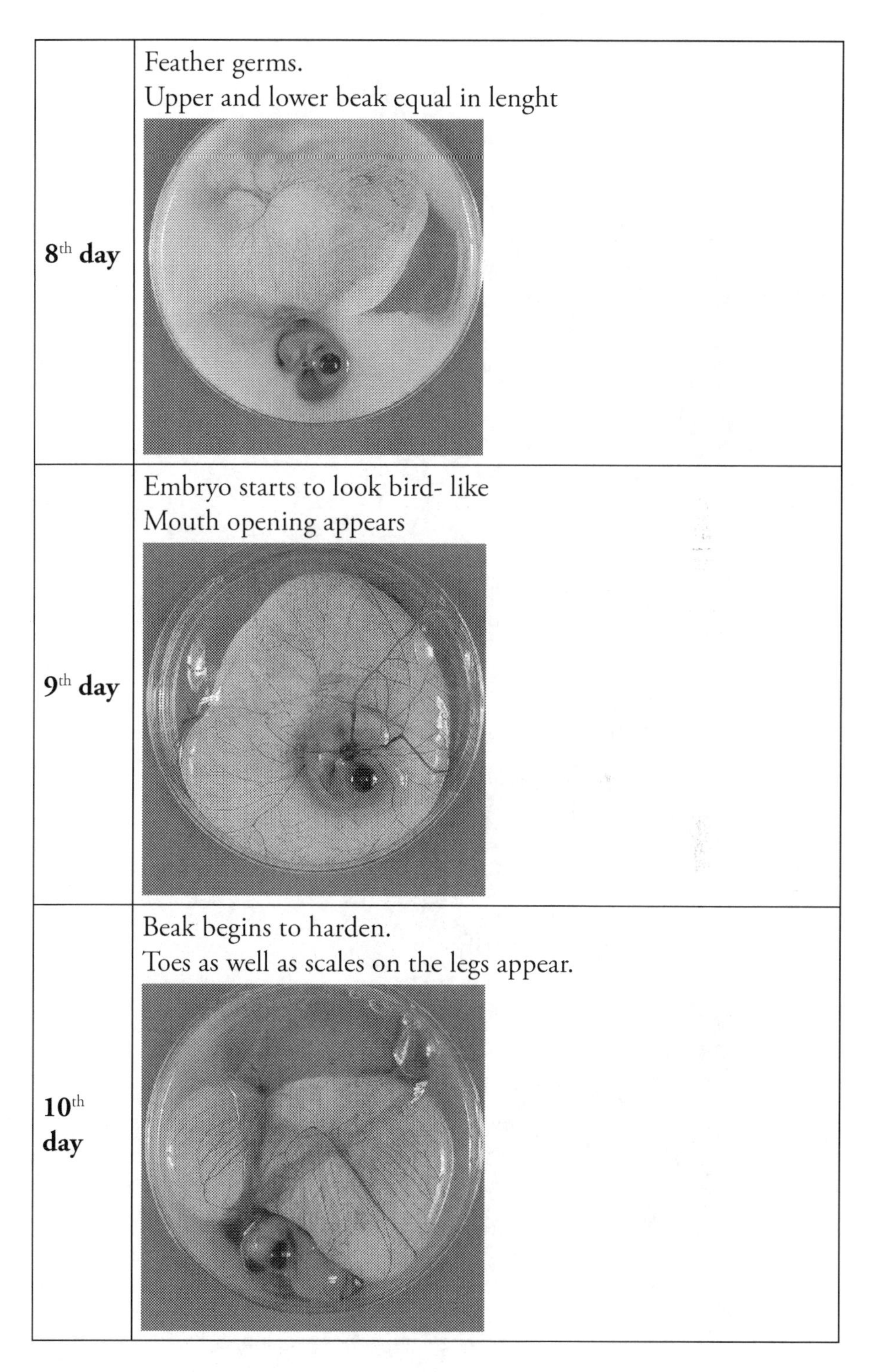

8th day	Feather germs. Upper and lower beak equal in lenght
9th day	Embryo starts to look bird- like Mouth opening appears
10th day	Beak begins to harden. Toes as well as scales on the legs appear.

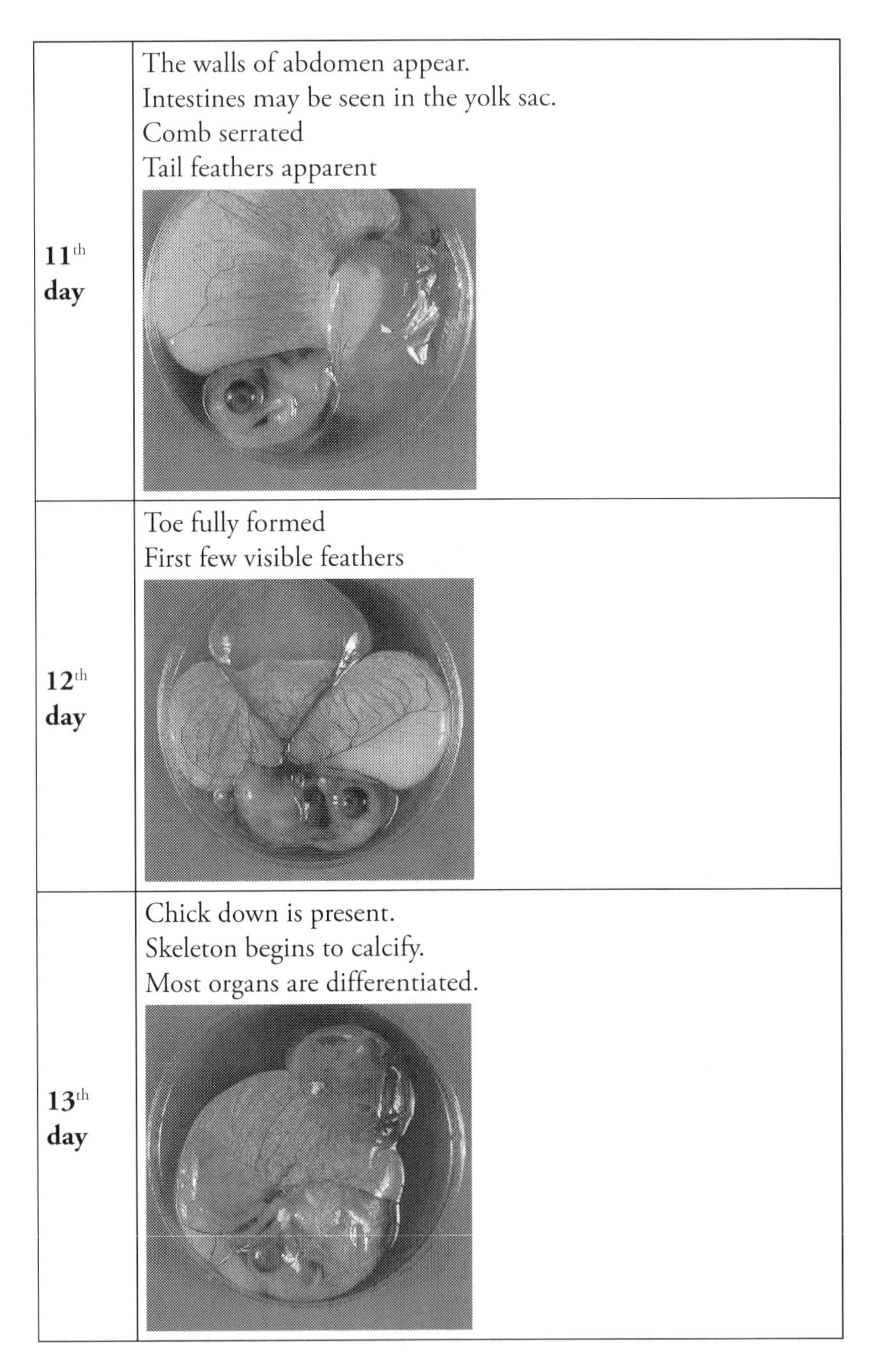

11th day	The walls of abdomen appear. Intestines may be seen in the yolk sac. Comb serrated Tail feathers apparent
12th day	Toe fully formed First few visible feathers
13th day	Chick down is present. Skeleton begins to calcify. Most organs are differentiated.

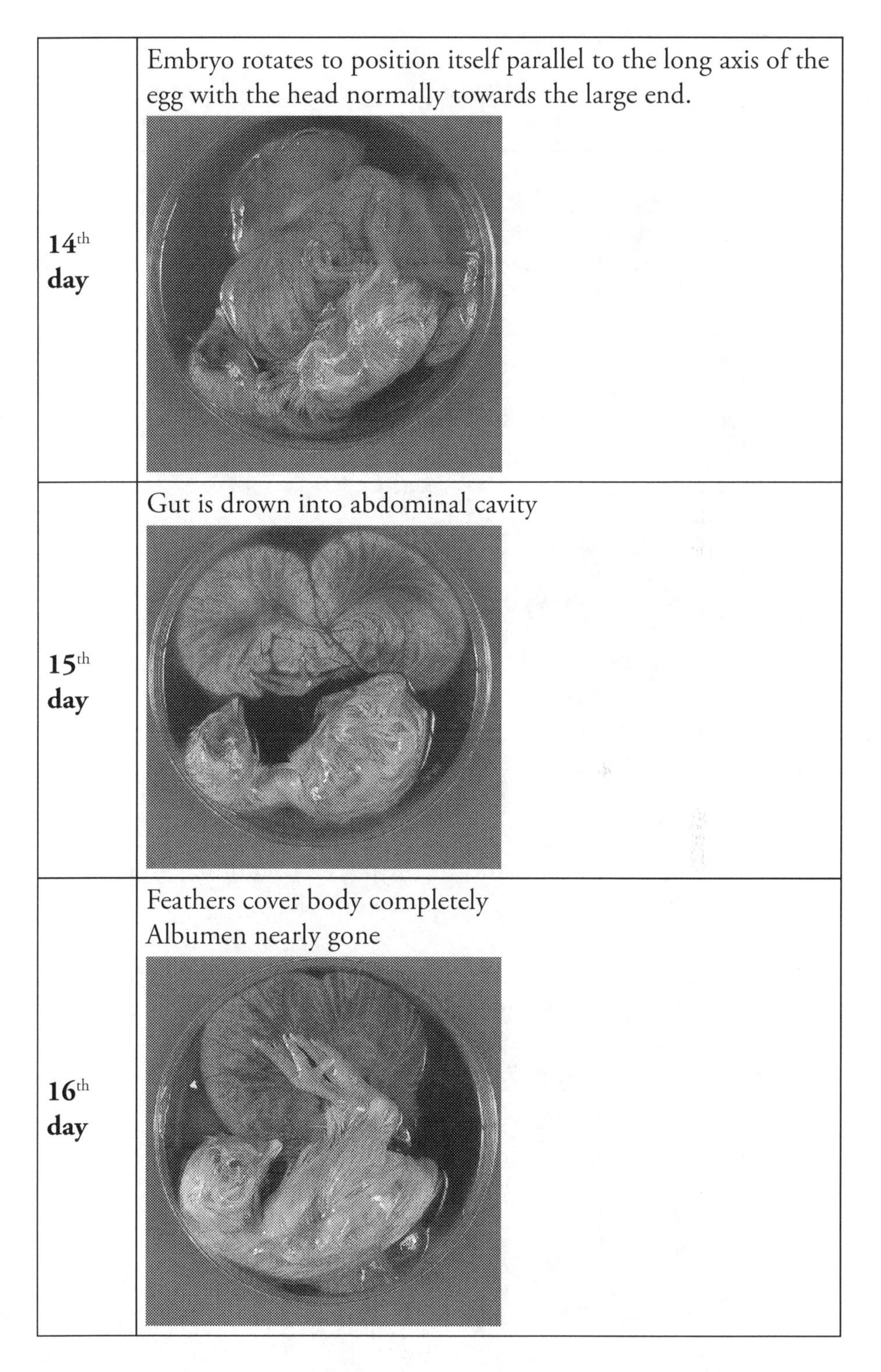

14th day	Embryo rotates to position itself parallel to the long axis of the egg with the head normally towards the large end.
15th day	Gut is drown into abdominal cavity
16th day	Feathers cover body completely Albumen nearly gone

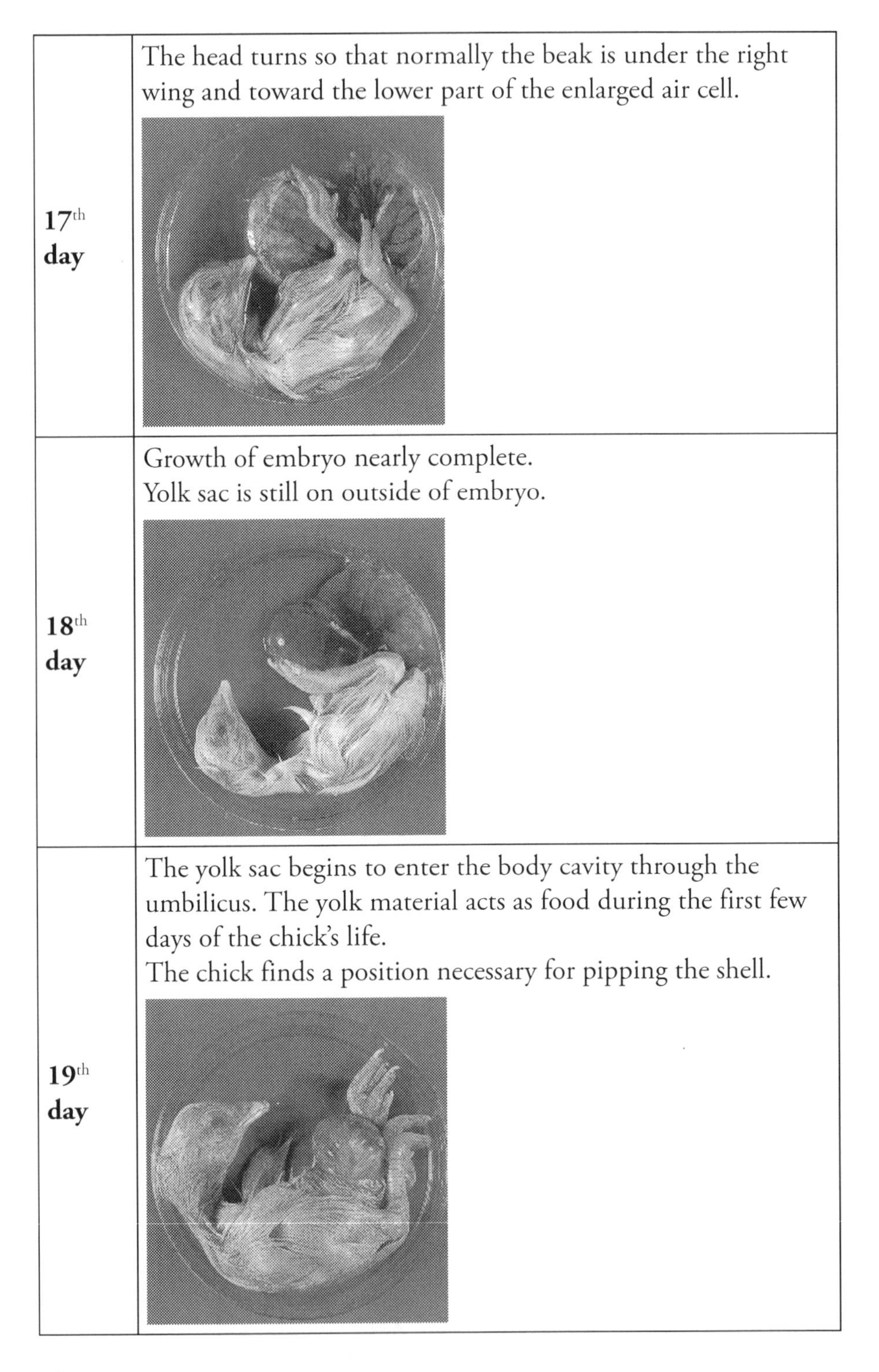

17th day	The head turns so that normally the beak is under the right wing and toward the lower part of the enlarged air cell.
18th day	Growth of embryo nearly complete. Yolk sac is still on outside of embryo.
19th day	The yolk sac begins to enter the body cavity through the umbilicus. The yolk material acts as food during the first few days of the chick's life. The chick finds a position necessary for pipping the shell.

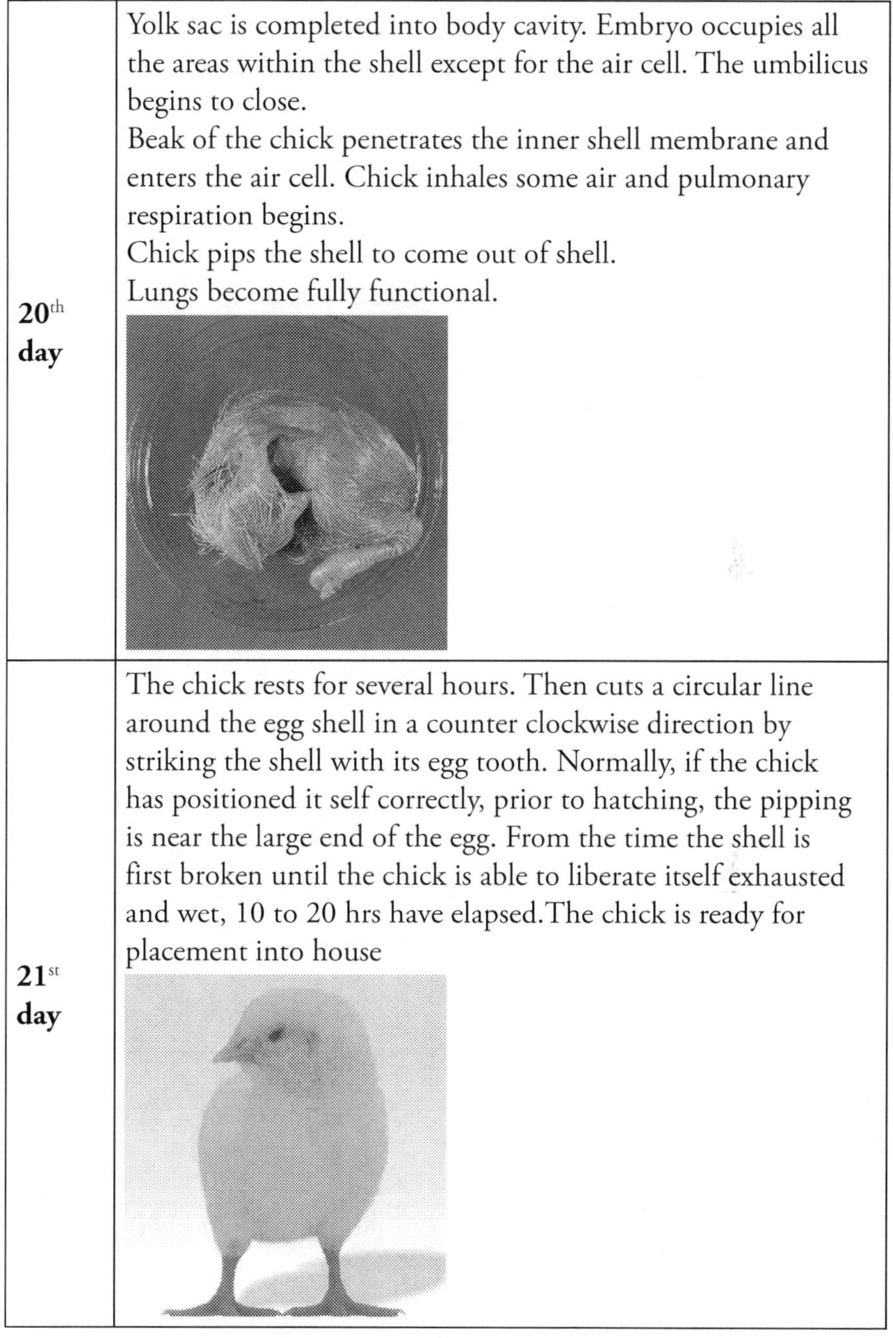

20th day	Yolk sac is completed into body cavity. Embryo occupies all the areas within the shell except for the air cell. The umbilicus begins to close. Beak of the chick penetrates the inner shell membrane and enters the air cell. Chick inhales some air and pulmonary respiration begins. Chick pips the shell to come out of shell. Lungs become fully functional.
21st day	The chick rests for several hours. Then cuts a circular line around the egg shell in a counter clockwise direction by striking the shell with its egg tooth. Normally, if the chick has positioned it self correctly, prior to hatching, the pipping is near the large end of the egg. From the time the shell is first broken until the chick is able to liberate itself exhausted and wet, 10 to 20 hrs have elapsed.The chick is ready for placement into house

Source:Cobb Vantress USDA-ARS, Athens Georgia

Pictures obtained from Cobb Vantress Hatchery Guide

6

Hatchery Design

Chick hatcheries are modern buildings that provide separate rooms for each hatchery operation. Good design of hatchery is essential for cost effective operation. Hatcheries form part of the food chain and their design must therefore, incorporate food hygienic standard.

6.1. DETERMINING THE SIZE OF HATCHERY:

The size of the hatchery is computed as follows:-

1. Egg capacity of the setters and hatchers.
2. Number of egg that can be set each week.
3. Number of chicks that can be hatched from each setting.
4. Number of chicks that can be hatched each week.

6.2. STRUCTURE:

Hatcheries should have the following features:

1. Durable wall and floor furnish and easy to clean drains. The wall surface should have a minimum of joints and fastenings that impede effective cleaning. A good floor finish can be obtained with cement incorporating hard stone aggregate or topped with a self leveling epoxy which has certain advantages over the more traditional finishes. The floor must be sloped to drains in each

room of the hatchery building. All drains need to be trapped particularly in hatching and pull areas to prevent blockages from egg shell and debris. The entire drainage system must be designed to handle large quantities of wash water and solid matter.

2. A biosecure flow of eggs, chicks and equipment through the building is essential. Clean and dirty area must be separated to prevent cross-contamination by fluff that can be carried around the hatchery on air currents or staff clothes. To maintain a disease free hatchery it is essential that all the people entering the premises shower and change into clean clothing in an adjoining room. They may leave only through this same room where they must change back their clothing's. The shower room must be carefully constructed so that those entering the hatchery may gain entrance only through the shower. There should be an area for disrobing and for dressing in clean working clothes.

6.3. HATCHERY DESIGN

A hatchery is a highly specialized building with unique requirements for construction and operation. An appropriate location of a hatchery affects the profitability of the entire operation as well as the overall performance of the hatchery.

The following points should be considered when designing a hatchery:-

- Available capital
- Hatchery site
- Production capacity
- Type of hatchery equipment

6.3.1. AVAILABLE CAPITAL:

Design and the construction of a hatchery building depends upon the type of construction material, capacity, equipment selection and the level of automation.

6.3.2. HATCHERY SITE:

Generally a hatchery should be located on high ground and should have a good drainage. In order to provide bio-security, the hatchery should be situated for enough away from the production units to minimize the spread of diseases. Hatchery must be located at least 1500 ft (460 m) away from the breeder farm. The availability and cost of labour, other utilities and waste disposal should be also considered.

6.3.3. PRODUCTION CAPACITY:

Production capacity to a greater extent determine the size and cost of hatchery. Therefore, the importance of determining production capacity early in the planning phase is essential.

6.3.4. TYPE OF HATCHERY EQUIPMENT:

Selection of proper equipment is critical for hatchery performance. The type of equipment selected has a major impact on the total cost of the project.

6.3.5. BASIC FLOOR PLANS FOR HATCHERIES

Generally two building for hatcheries are available as rectangular or T-shaped.

- Rectangular shaped buildings are often used for small hatcheries. The rooms are oriented into a clean to dirty sections. But this design is less suitable for expansion and lacks some bio-security features (Fig. 6.1).
- T-shaped buildings are most common ones. Here incubator and hatcher halls are located in each wing and are separated from storage and processing rooms by a corridor. This design gives the option of starting small by building only one wing with a centre area large enough to accommodate a future wing (Fig. 6.2 and 6.3).

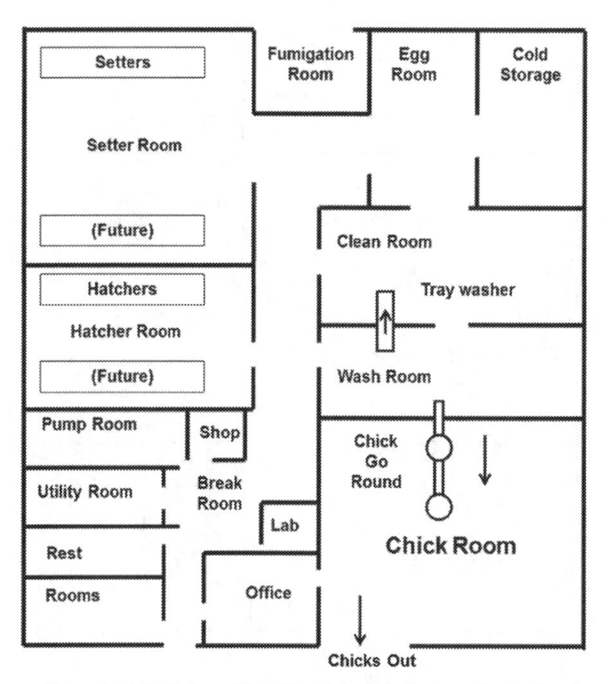

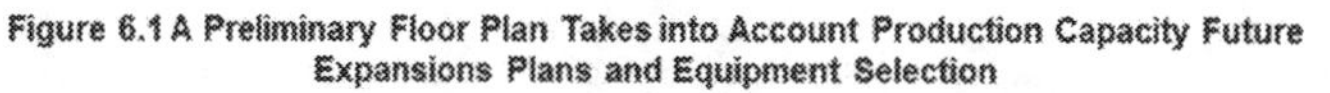

Figure 6.1 A Preliminary Floor Plan Takes into Account Production Capacity Future Expansions Plans and Equipment Selection

Source: Commercial Chicken , Meat & Egg Production

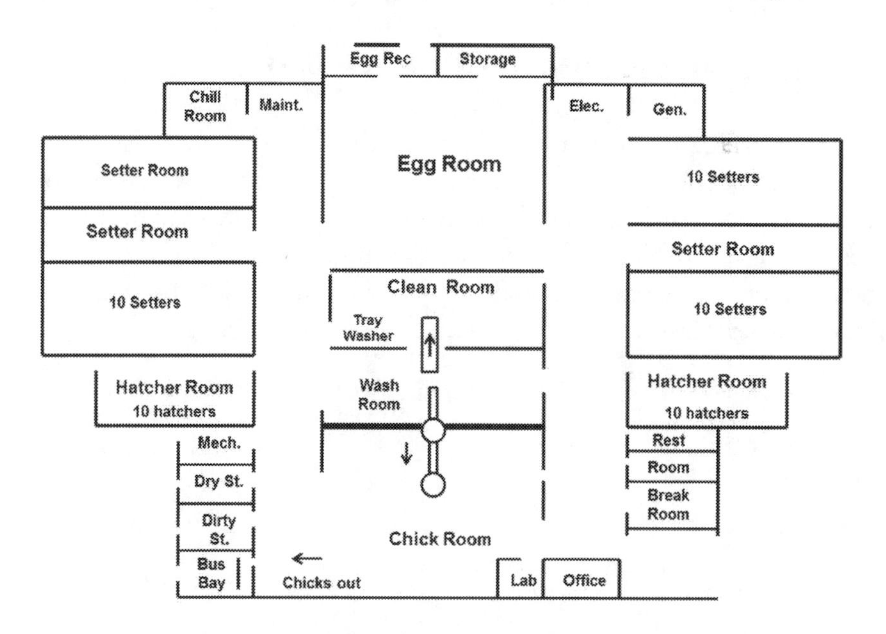

Figure 6.2 T-Shaped hatchery floor plan with 40 standard incubators and 20 hatchers, 46,555 square feet (4,324 square meters) capacity of hatchery: 1,267,200 eggs set per week (65.9 million per year).

Source: Commercial Chicken , Meat & Egg Production

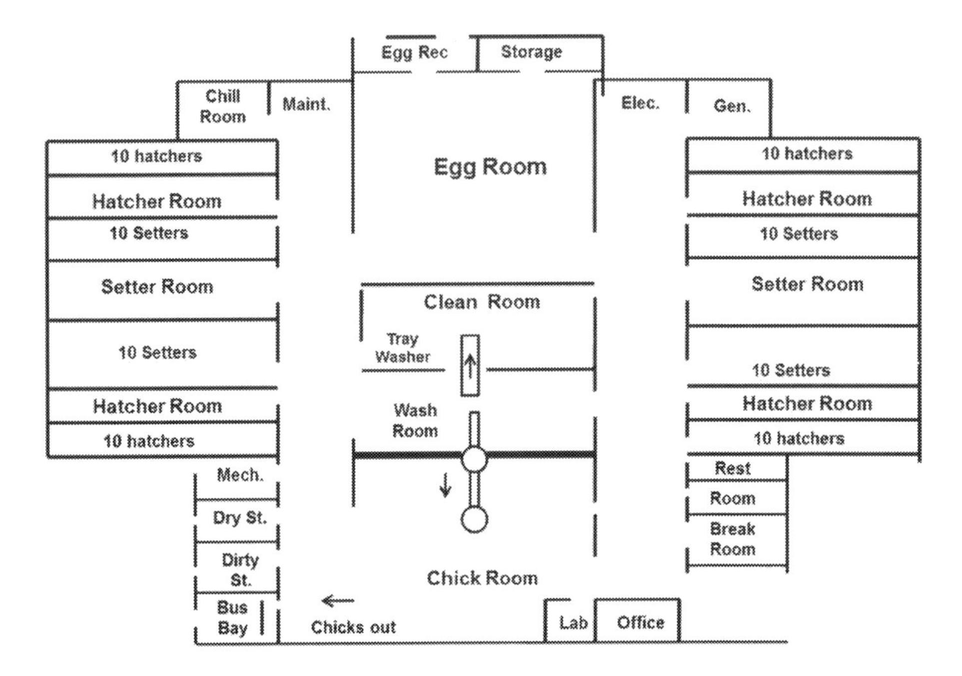

Figure 6.3 T-Shaped hatchery floor plan with 40 rear door incubators and 40 hatchers
Source: Commercial Chicken , Meat & Egg Production

6.4. EGG CHICK FLOW THROUGH HATCHERY:

Hatcheries should be constructed so that the hatching eggs may be taken in at one end and the chicks removed at the other. In other words, eggs and chicks should flow through the hatchery from one room to the next room. There should be no backtracking. Such a flow affords better isolation of the rooms and there is less human traffic throughout the building. A diagrammatic flow system is shown in Fig 6.4 and 6.5.

Fig -6.5: Egg Temperature Flowchart (for fresh eggs)

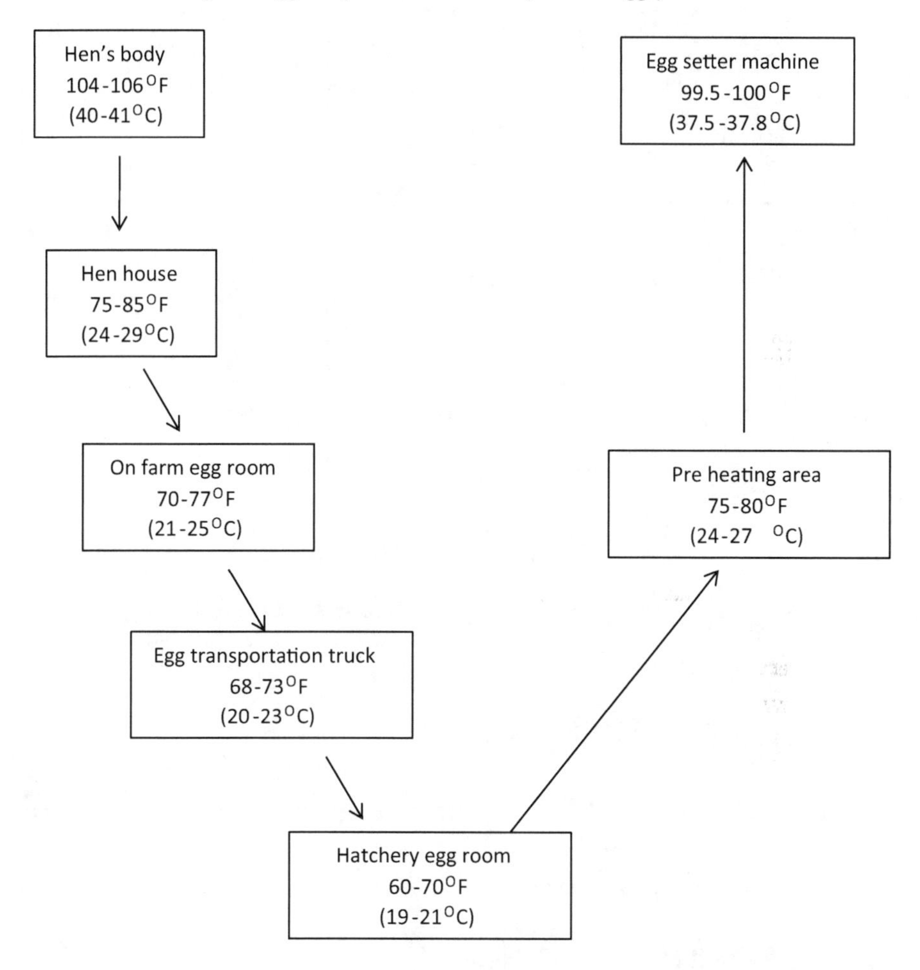

Source: Cobb Hatchery Management Guide

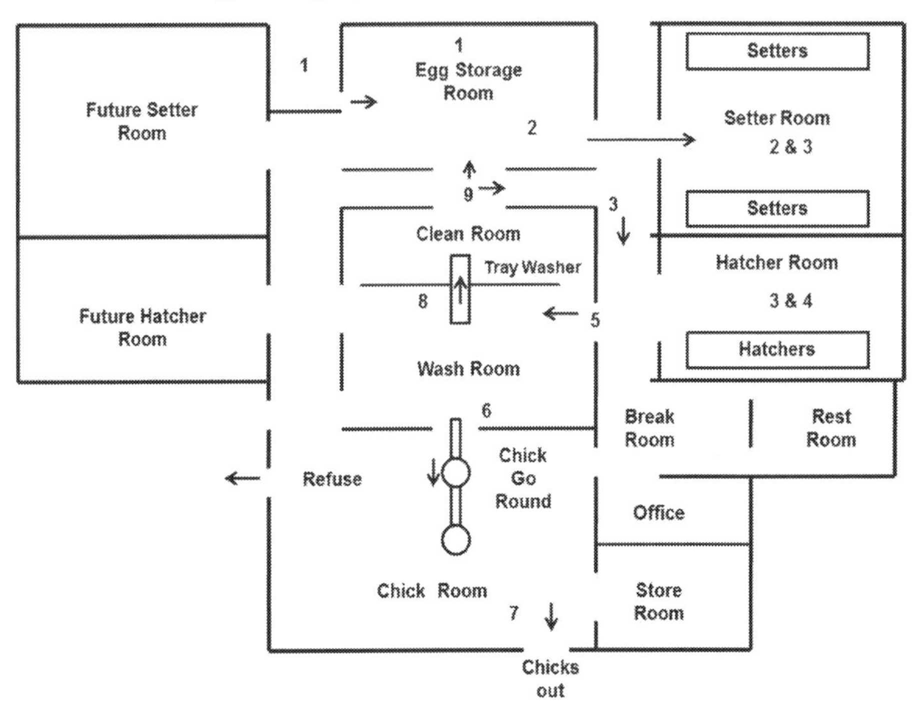

Figure 6.5 Typical hatchery Flow

Source: Commercial Chicken , Meat & Egg Production

6.4. HATCHERY BUILDING:

Hatchery building should be intricately designed, properly constructed and adequately ventilated. The brief general points for designing a hatchery building are as follows:-

6.4.1. SIZE OF THE BUILDING:

The width of setter and hatcher rooms will be determined by the type of incubators used. The total width of the rooms may be determined based on the depth of incubators, space for working aisles and the space needed behind the machines. The other rooms in the building may be designed in such a way as to provide a proper egg flow. The building have a proper roof which may be of different designs such as flat, gable, monitor or shed type. Most commercial hatcheries are built with forced draft ventilation. The recommended height of ceiling should be 10 feet. The walls should be constructed from fireproof material. Since the interior of the hatchery building is continuously being washed and disinfected, the inside walls should be covered with a glazed hard non-absorbent finish. These finishes prevent the growth of molds common to walls that are porous and absorbent. Concrete blocks lend themselves very well to wall construction. They are painted with a material that seals the pores of the blocks and leaves a hard glazed surface. The inside walls should be constructed from water proof material.

Most hatchery rooms have a high humidity and during cold weather condensation of moisture on the ceilings is common. Thus it is impractical to construct the ceilings of any material that will deteriorate in presence of water. Plaster should not be used. The best materials are water proof pressed wood or metals. Insulation of the ceiling will eliminate a good share of moisture condensation. Adequate ventilation in the rooms will also help as will negative pressure. All floors must be concrete, preferably with embedded steel to prevent cracking. The concrete must be given a glazed finish. There must be no uneven or spots on the floor. All water must drain away, none should be left standing. The floor drains should not have slope of 10 feet having no smaller than 6-inch trap type floor drains having a flat top so that hatchery equipment could be rolled over them. Gutters must be constructed about 6 inch wide and 6 inch deep in concrete floor in such a way that water will run to the end of the gutter and finally into the sewer system. Cover the gutter with a steel plate in which holes have been drilled or with slotted cast iron covers built for this purpose. Sewers should be larger than those used in most industrial buildings. Water lines should be placed below the concrete floor. Large amounts of water will be used in the hatchery for washing hatching trays and cleaning as well as in the incubators. Be sure the incoming water lines are adequate in size and the water pressure is 50PSI (3.51 kg/cm^2) in setters and hatchers. On an average 3028 liters of water per day will be used in the entire hatchery for each 100000 incubator egg capacity but the figure is highly variable depending on the hatchery equipment used.

6.4.2 OUTLAY OF DIFFERENT ROOMS INSIDE THE HATCHERY BUILDING

1. EGG RECEIVING ROOM:

The hatchery building must have adequate size of room to receive the daily collected hatching eggs from the farm.

2. EGG HOLDING ROOM:

Proper construction of the egg holding room is important as the quantity of the hatching eggs is to be maintained. It should be about 8 feet high insulated, slowly ventilated with complete air movement, cooled and humidified. The room should be refrigerated to maintain a temperature of 65°F (18.3^{O}C). A forced draft air type of refrigeration unit is required in order to keep a uniform temperature throughout the room. The capacity or size of the refrigeration unit is measured in Btu/hr which is the rate of heat removed. Sometimes the unit of measurement is in tons. A ton of refrigeration is equivalent to 12,000 Btu/hrs. The room insulation, outside temperature and the other factors will determine the size of the unneeded to adequately cool the egg-cooler room. Size and ratings of refrigeration units for egg holding rooms is presented in Table—6.1. The following calculation will provide an estimate of the size required Btu heat removed/hr.

1. Floor area in ft^2 x 3
2. Wall plus ceiling area in ft^2 x 4
3. Number of dozen eggs cooled/day x 5.5
4. Miscellaneous:

35 Btu for each 10 ft^2 of floor space.

Total Btu required = ___________________________

Refrigeration units vary in size and cooling capacity.

The following information can be used to establish actual room sizes and layout:-

6.5.1. Egg Room: The size of this room depends on how many eggs are received. When calculating egg room size, provide 4 sq.ft. of floor space for every 1000 eggs to be stored (0.372 sq.m.). The minimum egg room size should not be less than 600 sq.ft. (55.74 sq.m.). The height of ceiling should be 12 ft. (3.65 m.).

6.5.2. Fumigation Room: The fumigation room should be large enough to hold at least one-half of egg cases and buggies in a single day.

6.5.3. Prewarming Room: A floor space of 15 sq.ft. (1.4 sq.m.) for each egg buggy with enough air circulation around all eggs to maintain an even temperature.

6.5.4. Setter Room: Usually 24 to 30 inches (60-70 cm) between the ends and back of setters and the walls is considered adequate space for this purpose. The front setter should have a space at least 10 to 12 ft. (3.0 to 3.6 m.) wide from the front to a facing wall or a facing row of setters. A ceiling height for the setter room is 14 ft. (4.27 m.). This provides ample room for working and cleaning on top of the setters.

6.5.5. Hatcher room: Hatchers must have a front aisle of at least 10 ft. (3.05 m.) wide when two rows of hatchers face each other, a minimum aisle space of 10 to a 12 ft. (3.05 to 3.65 m.) should be permitted for working in both rows at the same time. Place hatchers 24 to 30 inches (60 to 70 cm) from end and back walls for cleaning purpose. A 14 ft. (4.27 m.) ceiling height should be provided for cleaning the tops of the hatchers.

6.5.6. Chick room: The size of the chick room is based on the maximum number of chicks processed daily. Whenever, full chick boxes are held in the chick room, always place them on dollies and stack them at least 12 inches (30 cm) apart to allow for proper ventilation. A minimum of 12 sq.ft. (1.12 sq.m.) per 1,000 chicks, upto 20 sq.ft. (1.86 sq.m.) per 1,000 chicks stored in the chick room. Usually a hatchery with a capacity of 1,663,200 eggs set per week on a 4 daily per week hatching schedule the chick room would have to handle approximately 3,41,000 chicks per hatch. Therefore, the chick room should be approximately 4,400 sq.ft. (409 sq.m.).

TABLE 6.1: SIZE AND RATINGS OF REFRIGERATION UNITS FOR EGG HOLDING ROOMS

Dozen of eggs/day	Room size		Refrigeration unit		
	ft	m	HP	Ton	BTU
800	11 x 12 x 7	3.3 x 3.7 x 2.1	½	½	6600
1200	12 x 21 x 7	3.7 x 6.4 x 2.1	¾	¾	9080
1600	16 x 21 x 7	4.9 x 6.4 x 2.1	1	1	12000
3100	21 x 28 x 7	6.4 x 8.5 x 2.1	2	2	24000

(Source: Hatchery Planning Company, Georgia, 30106, USA)

Egg holding room should have a 75% relative humidity to prevent eggs from drying too rapidly during the pre-incubation period.

3. SETTER/ HATCHER ROOMS:

These rooms must be adequate in size. It is better to have them too large than too small. Usually hatcheries of medium size will hatch chicks twice a week but large hatcheries will hatch as many as six times a week in order to equalize the daily work and equipment load. Consequently, hatching schedules will affect the size of the some rooms in the hatchery. Table 6.2, shows the general recommendations for floor space of the various hatchery rooms when there are two hatches per week.

TABLE 6.2: FLOOR SPACE FOR HATCHERY ROOMS (TWO HATCHES/WEEK)

Room type	Per 1000 Eggs Incubator-Hatcher		Care of Per Eggs Set Per Setting (360 Eggs/Care)		Per 1000 Straight Run Chicks Hatched Per Hatch	
	ft^2	m^2	ft^2	m^2	ft^2	m^2
Egg Receiving Room	2.0	0.19	4.32	0.40	15	1.39
Egg Storage Room	0.33	0.03	0.71	0.06	2.47	0.23
Chick Holding Room	4.0	0.37	8.64	0.80	30.0	2.79
Washroom	0.80	0.07	1.73	0.16	6.0	0.55
Storage Room	0.70	0.07	1.51	0.14	5.25	0.44

(Source: Hatchery Planning Company, Georgia, 30106, USA)

Setters and hatchers will operate more uniformly and economically if the relative of these two rooms is maintained at 50%. The size of these two rooms will depend on the make of equipment used. The manufacturer will be able to supply these dimension and other necessary construction details. The incubating equipment takes relatively little floor space.

4. FUMIGATING ROOM:

The fumigation room should be as small as possible in order to reduce the amount of fumigant used. It should be the size necessary to hold one pick up of eggs. A fan should be used to circulate the air and exhaust the fumigant.

5. WASHROOM:

Here the chicks are removed from the hatching trays and box an exhaust hood in this area to reduce cross contamination. A chicks are boxed they are moved to the chick holding room and t are washed in a tray washer in the washroom.

6. CHICK HOLDING ROOM:

After the chicks are hatched they are boxed and moved to tl holding room. Maintain a 65% relative humidity to prevent e chick dehydration

7. OTHER ROOMS:

Depending on the type of hatchery following other ro constructed:

1. Office
2. Small laboratory
3. Tool room
4. Generator room
5. Box storage room
6. Egg grading and traying room.

6.5. DETERMINING THE ROOM REQUIREMENTS

The single most important factor in establishing the size of the is the planned production capacity. This is important because of the egg holding room, chick room, box storage room and clea are predicated on chick production. Smaller hatcheries usuall two days per week, while larger broiler hatcheries may hatch fo days per week. Once the daily input of eggs and output of chi been established, the actual room sizes and layout of the hatcher developed.

6.5.7. Wash room: A floor space of 6 sq.ft. (0.56 sq.m.) per 1000 chicks hatched or 12 sq.ft. (1.12 sq.m.) times the total number of hatcher buggies per hatch day, plus adequate space for cleaning, take-off and automation equipment.

6.5.8. Clean tray room: A clean tray room should be sized for 15 sq.ft. (1.4 sq.m.) for each buggy stored on hatch day.

7

Hatchery Equipment

Good equipment plays an important part in increasing hatchery profits. Not only will the hatchability of the eggs and chick quality be improved but labour costs will be lowered. All are involved with an emphasis on the increase in the efficiency of the operation.

Hatchery equipments cannot be standardized because many factors are involved in the smooth running of these equipments as:-

- Size of the hatchery
- Number of hatches per week
- Whether breeder chicks or commercial chicks are hatched.
- Type of incubator.
- Type of disease control programme.

The following equipments are involved for running a hatchery:

7.1. INCUBATING EQUIPMENT:

Through the years the forced-draft incubator has undergone a variety of changes to make it a highly specialized equipment with lighter cabinet materials, floorless machines, easier cleaning, improved thermostats, automatic egg turnings, more accurate humidifiers and better cooling devices.

a. EGG SETTERS AND HATCHERS:

A setter is equipment in which the eggs are loaded for first 18 days of incubation period. Setters are available of varying capacity. A specific temperature is maintained throughout this period. There are proper devices for recording temperature and humidity with digital readout mounted in front of the machine. There is a control panel present on the top of front of machine comprising of primary and secondary heaters. Dry and wet bulb temperatures are read out on the front of the machines. Some show the relative humidity rather than the wet bulb thermometer reading. In order to maintain proper temperature and humidity high speed electric fans are fixed inside the cabinet of the machine. These fans can be removed in one piece for cleaning. Indicator lights are fixed on a panel at the front of the machine to show when power is on or off. For recording the turning in the machine a panel box is fixed on the side which records the number of turnings in 24 hours. Trays are racked one above the other inside the machine on a device which could be turned at an angle or stopped in any position during the period of 18 days.

Hatchers are available with varying capacities with an assembly containing fans, and heaters. The fans are fixed in the cabinet of the setters to create a more even temperature and humidity throughout the cabinet. Dry and wet bulb temperatures are read by large digital numbers on the front of the machine. Hatching machines lack the mechanism for turning. Plastic hatcher trays are stacked inside the hatcher to allow more air movement throughout the operation of the hatching.

b. EGG CANDLERS:

Eggs are candled during the incubation period. Eggs that are infertile and those in which the embryos have died are moved. Because of increasing hatchery size and increased cost of labour, considerable opportunities may exist for automating many of the labour intensive operations in hatcheries. As a broad guide, a staffing level of one employee per one million chicks per year is the norm without automation or one employee per two million chicks per year with automation. Machines are available to:-

- Grade eggs before setting.
- Candle and transfer eggs at 18 days.
- Perform in-ovo vaccination.
- Separate chicks from debris.
- Count chicks.
- Spray, vaccinate and box chicks,
- Remove debris.

A range of conveyers and elevators are available to speed up grading, sexing and other operations which need to be manually performed. Much of this equipment is precision made and very expensive and only very large hatcheries can justify its use. Nevertheless, smaller hatcheries can achieve benefits from equipment such as vaccum transfer machines and chick grading equipments which are inexpensive but deliver considerable productivity benefits. When selecting equipments, ensure that it can be disinfected easily, quickly and effectively. Egg and chick handling equipment should not contribute to cross contamination between eggs or between chicks.

c. EGG GRADING AND WASHING EQUIPMENT:

In many hatcheries it is necessary to grade eggs by size before they are set, in others, washing may sometimes be a practical procedure. Eggs are graded for size using automatic graders. There are many of these in the market of various types and capacities. One should be secured that will do the job adequately in a limited amount of time and with little egg breakage.

Hatching eggs can only be satisfactorily washed in commercial egg washers. The washing equipment should be able to hold eggs at 110°F (43 °C) for 8-12 hrs using a cleaning compound containing 200 ppm of chlorine.

d. VACCUM EGG LIFTS:

To expedite the removal of eggs from egg flats, vaccum egg lifters are used. Most of these lift the eggs by suction. The lifters are operated by a vaccum pump that creates a suction on rubber cups the number varying from 12-48. Most of those used in hatcheries are 6 x 6. So that they lift eggs from filler flats that are six row long and six row wide, lifting 36 eggs at a time. Some are larger (6 x 8), lifting 48 eggs. The size of the lifters is determined by the size of the egg flats used for collecting eggs and delivering them to the hatchery.

e. EGG CONVEYERS:

Reducing the cost of labour calls for many labour saving devices. Although hatching eggs are originally cased and delivered to the hatchery, this process necessitates that the eggs be handled several times. A more modern method is to place the eggs in the setting trays at the farm, then slide the trays into buggies and transfer them to the hatchery.

Several types of carts may be used to expedite their transfer as follows:-

I. WHEELED CARTS: These have four wheels, no platform and come in several sizes.
II. SEMILIFT CARTS: These are larger carts with two wheels and have a capacity of 20-25 dozen cases of eggs.
III. HAND TRUCKS: These light weight trucks are capable of handling four or five cases of eggs.
IV. PALLETS: Cases of eggs, egg cases and chick boxes may be placed on pallets to be moved later with a fork lift.

f. CHICK SEXING EQUIPMENT:

These are specialized sexing machines capable of visually observing the sexual organs of the chicks.

g. VACCUMS CLEANERS:

Hatchery dust should be vaccumed, not swept. Large industrial type vaccums should be used. The size and number will be determined by the size of the hatchery operation. There are many types of vaccum cleaners available in market as:-

1. SELF CONTAINED OR CONVERSION COVERED MACHINES: Self contained units are complete conversion units made to be placed on top of a steel barrel.
2. DRY OR WET AND DRY UNITS: Some vaccums pick up only dry material while as others are used with either wet or dry debris.
3. WITH OR WITHOUT COLLECTORS: Dust collectors are usually bags that move with the machine and are easily cleaned.
4. PORTABLE HAND UNITS: These are either smaller vacuums that may be carried or those with smaller suction heads. These are helpful in cleaning the inside of setters and hatchers.

h. PRESSURE PUMPS:

Increased water pressure is necessary to do a thorough job of washing the floors, walls, incubators and hatching trays. A pressure pump should be installed in the hatchery to give the necessary water pressure. These come in various sizes and capacities, some are portable and some must be permanently installed.

i. TEST THERMOMETERS:

Several highly accurate test thermometers should be a part of the hatchery equipment. These may be used for checking the accuracy of the thermometers in the setters and hatchers.

j. COOLERS:

During periods of hot weather it becomes necessary to cool the hatchery building. Certain rooms require more cooling than other. Chick room is usually the first room to show temperature increase because of the build up of heat from the chicks followed by the hatching room. The most economical method of reducing the temperature in hatchery building is by evaporative cooling. The theory of evaporative cooling is based on the fact that when water evaporates a cooling effect is produced.

A practical application of the above phenomenon is through the use of commercial evaporator-coolers. These come in varying sizes from 2 to 20 feet on a side. The larger the size, the more air moved. In the evaporator cooler moisture is provided by the moisture laden pad. Air is sucked through the pad, usually by a squirrel-cage fan moisture is absorbed, the air is cooled and forced throughout the building in ducts, then out of the rooms through an exhaust opening. A more common method is to use exhaust fans. To maintain equilibrium in the room between incoming and outgoing air, the capacity of the exhaust fans should be about 10% greater than that of the intake fans, thus creating negative pressure. Generally smaller coolers side of the building and the air blown directly into small rooms. In case of larger installations the coolers are fixed at or near the peak of the roof. The specifications for evaporative coolers are given in the Table: 7.1.

TABLE 7.1: EVAPORATOR—COOLER SPECIFICATIONS

Width (inch)	Depth (inch)	Height (inch)	Outlet Velocity (ft/min.)	Air Delivery (ft^3/min.)
27	27	25	1300	2300
34	28	30	1750	3100
34	42	42	2500	5600
38	45	42	2400	7700
56	56	60	1980	9100

(Source: Hatchery Planning Company, Georgia, 30106, USA)

8

Hatchery Management

Incubation is a process of 21 days during which the domestic chicken; a microscopic germ is changed into a chick. It can also be defined as management of fertile eggs to ensure satisfactory development of germ/embryo inside egg and production of normal and healthy chicks. The eggs produced by fertilization of ova with sperm are called fertilized or hatching eggs. These eggs are produced by the rearing the laying hens along with cock in a proportion of at least 10:1 or with the help of AI. However, laying hen can produce eggs naturally without mating with male. This type of egg does not contain embryo and is called unfertile egg.

8.1. METHODS OF INCUBATION

Hatching of selected eggs can be done by incubation methods and there are two methods of incubation as:-

1. NATURAL INCUBATION:

In this method eggs are incubated with the help of a broody hen. It is a primitive method of hatching eggs. This method is still popular with small poultry keepers in remote rural area and peri-urban areas. The selection of hatching hen should be done based on health status of hen having no deformities with anatomical good size. The hen should neither be very heavy nor very small. The hen should have good feathering. The behavior of the bird should be good for hatching, broody characteristics, quite, good sitter. The hen should be tested with the help of dummy eggs for her interest to sit on eggs. After selecting the hen, she should be treated for external and internal parasites. The nest should be made of fine, soft hay, straw or leaves and placed near the ground so that the hen could enter it without difficulty. The best time to set is at night because she is more likely to settle down to her job and hatching will also be on the night of 21st day and will have the whole night to rest and gain strength.

2. ARTIFICIAL INCUBATION:

Artificial hatching operation can be divided into following main headings:-

a. BEFORE HATCHING OPERATIONS:

COLLECTION AND SELECTION OF FERTILE EGGS: It is better to select the better quality eggs for incubation from breeders that are:-

1. Well developed, mature and healthy.
2. Compatible with their mates and produce a high percentage of fertile eggs.
3. Are not disturbed much during mating season.
4. Fed a complete breeder diet.
5. Not directly related.

Select the eggs by avoiding excessively large or small (50-55 gms: Chicken; 65-70gms: Ducks and 80-85 gms: Turkey). Cracked or thin shells, misshapen and dirty eggs should be avoided.

- Proportion of white to yolk should be 2:1.
- The shape of egg should be oval.
- The eggs should be free from blood and meat spots.
- The eggs should be collected from disease free stock.

8.2. TYPES OF INCUBATORS:

The size and type of incubators selected depends on the need and future plans of each producer. For continuous settings, separate incubator and hatcher units are recommended. If all eggs in the unit are at the same stage of incubation, a single unit can be used. It is essential that the room has a good ventilation system to supply plenty of fresh air.

There are basically two types of incubators available. They are as:-

- Forced-air incubators.
- Still-air incubators.
- **FORCED-AIR INCUBATORS** have fans that provide internal air circulation.
- **STILL-AIR INCUBATORS** are usually small without fans for air circulation; air exchange is attained by the rise and escape of warm, stale air and the entry of cooler fresh air near the base of the incubator. The physical conditions like desired level of temperature, humidity, ventilation and frequent turning of eggs are essential for proper incubation of eggs irrespective of method of incubation, these conditions are created by broody hens and under artificial methods incubators are operated either manually or by using automatic devices to maintain these conditions.

1. **TEMPERATURE:** Temperature appears to have direct effect on duration of hatch, size of embryo, embryonic mortality and viability of chicks after hatching. But the extent of adverse effect of abnormal temperature on quantity and quality of chicks depends upon extent and duration of deviated temperature from the optimum and also on the stage of incubation when such situation has occurred. Species of birds vary in their temperature requirement for incubation. The temperature in incubator and hatchery should be watched even after the pilot light goes off because growing embryos become exothermic after 13 days or so and the heat generated by embryonated eggs themselves raises the temperature beyond critical level. The sub-optimal incubation temperature causes late hatch and poor hatchability.

OPTIMUM TEMPERATURES FOR HATCHING EGGS: Embryonic growth may be divided into three phases, each requiring a different as follow:-

a. **PRIOR TO EGG LAYING:** The body temperature of the broody hen fluctuates between 105° and 107 ^{O} F (40.6 ^{O} to 41.7 ^{O}C). As the new embryo completes many cellular divisions during the 22 hours between the time of the union of the sperm and egg, cell and the time the egg is laid, the optimum temperature for embryonic development during this period must be that of the body temperature of the hen.

b. **DURING THE FIRST 19 DAYS OF INCUBATION:** Although varying slightly according to the make of forced-draft incubator the temperature is about 99.5 ^{O}F (36.7 ^{O}C).

c. **DURING THE 20TH AND 21ST DAYS OF INCUBATION:** In forced draft incubators best hatchability occurs when the temperature is lowered from that during the first 19 days to a range between 98 ^{O}F and 99 ^{O}F (36.7 ^{O}C and 37.2 ^{O}C). The final adjustments in the optimum temperature in the incubators vary according to the type and size of hatching egg under varying weather and other environmental conditions and adjustments made in incubating temperature accordingly. Under natural conditions, hens leave the nest many times each day during the incubation period. Cooling the egg during the hen's short absence from the nest is evidently not detrimental to hatchability under such conditions. During the artificial incubation there if the electric failures occur, they will cause a reduction in the environmental temperature within the incubator ranging from a short period of time to several hours. In this regard embryos are quite resistant to low temperatures during first 3 days of incubation but afterwards a linear reduction in the resistance to temperatures after 3 days upto 19 days. Hatched chicks in the hatcher are more resistant cold than to heat.

EFFECTS OF LOW INCUBATION TEMPERATURES: The following effects can be observed due to the low incubation temperatures:-

- Low temperatures lengthen the incubation period.
- Low temperatures increase the embryonic malpositions particularly during the first 19 days.

2. **HUMIDITY:** During incubation, the hatching eggs must lose certain proportion of moisture in order to lose some weight so as to produce strong and viable chicks. Chicken egg loses weight to about 10.5% through 19 days of incubation, thus the egg must lose about 0.6% of its original weight per day. Lower weight loss in large sized chicks and greater loss of moisture will produce small chicks than normal. When the eggs do not evaporate fast enough the chick will be larger than

normal. In either case the embryo is weakened, resulting in lowered hatchability and reduced chick quality. To regulate the evaporation of the egg contents the amount of moisture in the air surrounding the egg must be controlled, since this outside moisture determines the rate of the egg weight loss. High humidity reduces egg evaporation whereas the low humidity increases it.

Such RH is maintained by keeping the wet bulb reading of 88°F on 19th day and 90-91.5^{O}F on 20th and 21st day. The most accurate way to determine the correct humidity profile in the incubator is to look at the loss of moisture from the egg. When the weight loss is optimum, about one third of the shell will be removed after pipping.

- When the humidity is too high, the shell will be pipped nearer the blunt end of the egg.
- When the humidity is too low, the shell will be pipped towards the equator of the egg.

IMPORTANCE OF CORRECT HUMIDITY: In order to prevent dehydration of the egg contents, the relative humidity of the air in the setter during the first 19 days of incubation must be maintained within a narrow range of 50-60%. Increasing the humidity of the setter (1-19 days) lengthens the incubation period while as opposite results are obtained when the humidity is reduced.

MEASUREMENT OF RELATIVE HUMIDITY OF THE AIR: RH may be calculated by comparing the temperatures recorded by wet bulb and dry-bulb thermometers which are suspended in the same region. The dry bulb records the normal temperature of the air while as the bulb of the wet-bulb thermometer is encased in damped material and is chilled by the evaporation of the water. Generally greater the difference between the two thermometer readings, the less will be the relative humidity at a given dry bulb temperature. At the other end of the humidity range, when the air is completely saturated, can be no further evaporation from the material on the wet bulb and so both the thermometer will show the same reading. Annexure I, give a range of dry and wet bulb readings associated with specimen relative humidities that may be used in incubation and hatching egg storage.

TEMPERATURE AND HUMIDITY IN THE SETTER: There is an interaction between temperature and humidity in the setter. As the temperature has usually been set by the manufacturer, the only adjustment necessary is in the relative humidity whether the humidity is correct or incorrect depends only on the moisture loss through the eggshell by evaporation. A calculation should be made at least three timed during the incubation period because humidity varies from changes in egg weight, shell thickness, shell quality, age of breeding flock, season of the year and nutrition.

TEMPERATURE AND HUMIDITY IN THE HATCHER: During the last 2 days of incubation (20 and 21), while the eggs are in the hatcher, the humidity must be increased but only within certain limits. Correct moisture prevents the beak of the chick from sticking to the newly pipped shell and allows free movement of the chick's head at the time of pipping. Low moisture at the time of hatching will produce chicks smeared with egg or shell, stuck down and partially dehydrated. High humidity during this period will cause the chicks to be smeared with egg and having unhealed navals.

The relative humidity of the hatcher should be about 65% when the eggs are transferred to it, an increase of 5-10% from the setter. A temperature reduction of 0.5 to 1.5 ^{O}F (0.3 to 1.0 ^{O}C) is necessary. Generally a temperature reduction of 1.0 ^{O}F (0.6 ^{O}C) will decrease the relative humidity by 2.5%.

3. **POSITION OF THE EGGS DURING INCUBATION:** The ideal position for an egg during incubation is either quite flat, with the long axis parallel to the tray or else with the broad end higher than the narrow end. In either case there is the maximum room for the full development of the embryo and the best exposure of the shell around the air space to allow respiration. Eggs incubated with narrow end up show an unduly high proportion of dead-in-shell through failure of the embryo to establish access to the air space in time to allow pulmonary respiration to begin and so prevent the loss of blood from the ruptured allantois. In most cabinet-type incubators the eggs are trayed vertically with the broad end uppermost; through some trays carry the eggs flat. In flat type incubators the eggs are trayed flat.

4. **TURNING EGGS DURING INCUBATION:** Turning or altering the position of the eggs during incubation has definite influence on embryo mortality rate. It is necessary to ensure that the embryo is gently but frequently moved within the egg to prevent it setting and

adhering to other structures, as it would do if left for 21 days in one position. In the flat of incubator it also ensures more even heating of the egg contents, a factor that does not arise in the more uniform heating of the cabinet-type machine.

The need for frequent turning appears to be greatest in the earlier stages of incubation before the full development of the extra embryonic sacs and their fluids but it is none the less essential throughout the first 18 days in the incubator. No turning is advisable for the last three days of incubation. The interval between turnings can be as short as quarter of an hour provided that the eggs are turned in opposite directions each time, if the turning is always in the same direction as it seems to interfere with the centering action of chalzae and there is high embryonic mortality. In actual practice it is not necessary to turn as often as this but it is necessary to arrange the turning on a regular basis. In most of incubators turnings are entirely automatically controlled and the interval may be from ¼ to 4 hrs. In the cabinet machines trays are placed in vertical position and are turned through 40 to 45° each side of horizontal. It is important to avoid disturbance to the eggs in the first 24 hrs of incubation when the proper development of vitelline blood vessels from the blood islands can be inhibited. This would result in about 60% of embryos dying in the second and third days and a very low hatch of the others.

5. **VENTILATION:** Apart from the very early stages of development, the embryo is dependent for its supply of oxygen from the air that surrounds it in the incubator. It is important that the carbon dioxide and other gases produced by metabolism are removed from the vicinity of the egg so that they never reach a dangerous level and damage the balance of gas exchange. Ventilation of the incubator has dual purpose of providing the necessary oxygen and of removing excess carbon dioxide at a sufficient rate.

   An ordinary 57g egg needs about 5 lit. oxygen to enable the embryo to mature and at the same time it gives off about 3 litres of carbon dioxide. The oxygen content of fresh air is around 21% which is also optimum concentration for the development of hatching egg. It has been shown that the oxygen level can be reduced to 17.5% without affecting hatchability but reduction upto 15% reduces embryo survival. As long as there is continuous intake of fresh air and expulsion of used air, there is no risk of oxygen starvation under normal atmospheric conditions. For this reason, it is vital to make sure that the incubator room is thoroughly well ventilated, since it is the source from which the incubator draws its air.

a. **OXYGEN:** embryos use oxygen for their metabolism. The best hatchability is obtained with 1% oxygen level inside the incubator. Oxygen concentration above 21% may reduce hatchability but the embryo seems to be tolerant to excess oxygen than its deficiency. Approximately 21% of the air at the sea level is oxygen and it is impossible to increase the percentage appreciably in incubators unless pure oxygen is introduced. Although the oxygen content of the air in a commercial incubator is not altered, there may be some variation in the hatcher where large amounts of carbon dioxide are being liberated by the newly hatched chicks. In such cases, hatchability drops about 5% for each 1% that the oxygen content of the air drops below 21%.
b. **CARBON DIOXIDE:** Carbon dioxide is a natural by-product of metabolic process during embryonic development beginning during gastrulation. Normally fresh air contains 0.03-0.8% carbon dioxide depending on type of sample taken. In a badly ventilated room the level may be as high as 0-4%. For normal embryonic development the surrounding atmosphere should contain 0.4% carbon dioxide. Hatchability is markedly reduced at concentrations exceeding one percent and at 2%. There is no chance of survival as the carbon dioxide produced in the egg cannot be eliminated but the embryo dies due to accumulation of this gas in the blood circulation. To keep the carbon dioxide content within the safety zone it would require about 12 changes of air/hour in the incubator on the last days if all the eggs were at the same stage of development but in practice since the eggs are usually at three different stages, eight changes/hour are enough. Carbon dioxide concentrations in the air within the setter and hatcher when there is insufficient air exchange to remove it. Young embryos have a lower tolerance level to CO2 than old ones when measured as the concentration in the air within the machines. The tolerance level seems to be linear from the first day of incubation through the 21st day. During the first 4 days in the setter, the tolerance level of CO2 is 0.3%.

Carbon dioxide level above 0.3% in the setter reduces the hatchability with the significant reduction at 1% and completely lethal at 5.0%. Chicks hatching in the hatcher give off more Co2 than embryos contained in eggs and the tolerance level in the hatcher has been set at 0.75%. Recording devices are available for measuring the Co2 content of the air and some incubators have an inbuilt recording device. An adequate exhaust system connected to the hatcher should make sure that no harmful high levels of Co2 are reached in the machine. A good constant supply of fresh air is necessary to maintain the right level of Co2. Normal level of Co2 for the hatcher is 0.6-0.8% by volume.

NECESSITY OF VENTILATION IN HATCHERY:

Ventilation is required to:-

1. Supply oxygen
2. Remove carbon dioxide
3. Remove heat from the incubator
4. Provide incubators with air of proper texture. Most setters operate with a relative humidity of 50-60%, but the relative humidity of air of the incubator room will be quickly reduced once it enters the incubator where the temperature is higher and humidifiers will be needed in the machines to restore the moisture content of the air. Room relative humidity should be kept at about 50% outside air will often vary greatly from this percentage and must be conditioned prior to entering the incubator. Air entering the machines should have temperature of about 75°F (24°C). When the outside air is colder or warmer it will have to be heated or cooled in the incubator room prior to entering the incubators.
5. Remove heat produced in the hatcher and chick rooms. Each newly hatched chick will produce at least 2 Btu of heat per hour most of which must be removed from the building by the ventilating system. This heat production compares with about 30 Btu per hour for a 1.4 kg bird.

HATCHERY VENTILATION

Forced air should be used to ventilate the hatchery but each room must be considered a separate entity depending upon requirements for temperature, humidity and air.

Each room should be ventilated as a separate unit with displaced air exhausted outside the building. The requirement for fresh air in different sections of hatchery is given in Table 8.1.

TABLE 8.1: AIRFLOW PER MINUTE THROUGH HATCHERY ROOMS

Outside Temperature		100 Eggs			1000 Chicks
°F	°C	Egg Holding Room	Setting Room	Hatching Room	Chick Holding Room
10	-12.2	2.00	7.00	15.00	30.0
40	4.4	2.00	8.00	17.00	40.0
70	21.1	2.00	10.00	20.00	50.0
100	37.8	2.00	12.00	25.00	60.0

(Source: Pas Reform, P.O. Box 2, Zeddam, 7038ZG, Holland)

Incoming air should be heated in the winter and cooled in the summer. It should be humidified if necessary. More air should be moved through the rooms during summer seasons than winter season. Therefore, regulators should be installed on all ventilating fans to provide increased or decreased airflow to help control room temperature.

Setters normally draw fresh air from the room in which they are situated. This fresh air supplies oxygen and moisture to maintain the correct relative humidity. Air leaving the setter removes carbon dioxide and excess heat produced by the eggs. The air supply to the setter room should be 8 cfm (13.52 cubic meter/hr).

All setters have a humidity source that can control various levels of relative humidity. The fresh air supplies relatively little moisture and so to reduce the load on the internal humidification system, air entering the machines is pre-humidified to closely matches the internal relative humidity. The temperature of this air should be 76-80°F (24-27°C). Multi-stage setters require a constant amount of air. It should be adjusted so that the carbon dioxide level within the machine does not exceed 0.4%. Most fixed rack machines run at 0.2-0.3 % and buggy machines 0.3-0.4% but these elevated CO_2 levels are not required.

TYPES OF VENTILATING SYSTEM:

Positive air pressure is created in a room where the volume of air coming in is greater than that going out with the reverse; a negative room pressure is created. Positive or negative pressure in a room can often alter the operation of the ventilating system within the incubator. Therefore some incubator manufacturers specify positive pressure in the room; others negative pressure. But regardless the pressure inside the room and inside the machines will be the same. The exact room pressure will of course be affected by the air pressure outside the building but the increase or decrease of room pressure should never be greater than 1/8 inch static water pressure. Devices may be made that will regulate and record the static pressure.

9

Hatchery Borne Diseases

There are a number of diseases affecting the parent breeder flock thereby having an effect on the developing embryo, hatchability and chick quality. Many of the pathogenic organisms produce similar conditions causing high embryonic mortality, weak chicks and white diarrhoea. The important diseases affecting the eggs and the chicks produced in a hatchery are discussed here under.

9.1. PULLORUM DISEASE

Etiology: The disease is caused by a non-motile organism called Salmonella pullorum. It is a highly contagious disease and organism localizes itself in the ovary, liver, heart, testes and other body organs. The disease is widespread and unless precautionary measures are provided to control it, mortality will be high.

Host: Poultry, turkeys, quails, ducks can suffer from the disease. Broilers are more susceptible to the disease. It is a disease of newly hatched chicks. The chicks die between 2 and 7 days of age if they have infected during hatching and if they get infection after hatching, they show signs about 10 days post infection, upto the age of 3 weeks.

Symptoms: The newly hatched chicks may die without showing any marked symptoms. Others show symptoms similar to those of paratyphoid, such as huddling near the source of heat, somnolence, loss of appetite and white diarrhoea. Sometimes the adults may show symptoms similar to those of fowl typhoid.

Transmission: Infection to chicks comes from the infected eggs laid by a carrier hen. In the incubator the hatched, diseased chick spreads infection to the healthy chicks. Infection may enter through water or feed, contaminated with faeces of a diseased chick or also by inhalation of dust having bacteria, rarely through the eyes and wounds. There are asymptomatic carriers among adult birds. Infected hens may lay upto 34% infected eggs. Chick sexing can also spread infection from one chick to another.

Mortality: Maximum mortality occurs in 1st week and it declines in 2nd, 3rd & 4th week and may stop in 5th week of age. Mortality may range below 50% and rarely upto 100 percent. Approximately 5 to 35 percent of the surviving birds may become carriers.

PM Lesions: The newly hatched chicks which die from the disease may not show any gross lesions. Some of them may show haemorrhagic streaks on normally yellowish liver of a newly hatched chick. The chicks which die later may show grayish necrotic spots of 1 to 2 mm size on the liver, raised white spots on heart & spleen. The caeca may have semi-solid cheesy material inside. Unabsorbed and coagulated egg yolk is another important lesion.

Diagnosis: The *S. pullorum* organisms are easily isolated and identified in the laboratory. Cultures taken from such organs as the ovary, testicles, heart, liver and spleen are used to make the laboratory determination.

Once a bird has been infected with the *S. pullorum* organism, antibodies specific for *S. pullorum* bacteria make their appearance in the blood. These antibodies clump with the *S. pullorum* bacterial cells, inactivating them. A similar artificial reaction is used outside the body using agglutination test using whole blood mixed with an antigen, a specially prepared and standardized mixture of killed *S. pullorum* bacterial cells, dyes and solvents.

The following tests are conducted to diagnose the disease:-

a) Rapid-whole blood test.
b) Tube agglutination test.

As a general rule all breeder flock should be blood tested for pullorum disease. If the agglutinin test shows that certain birds are positive, these birds are to be taken to a diagnostic lab for confirmation of the test.

Prevention and Control:

1) Chicks should be obtained from breeder flock known and tested to be free from the disease as far as possible.
2) Do not mix new birds without first conducting agglutination test with antigen.
3) Do not feed leftovers from incubator, discard eggs or leftover human food etc.
4) The positive reactors should be immediately removed from the farm and destroyed.
5) Disinfection or fumigation of incubator and eggs must be routinely done.
6) The infected equipment must be thoroughly washed and scrubbed at least 3 times with 2% hot caustic soda. Fumigation of premises should also be done.

Treatment: Treatment does not prevent the birds from becoming carriers. The treatment of birds therefore, is a questionable proposition. For treatment, nitrofuran and sulphamirazin are good. Trimethoprim and sulphadiazine, ampicillin and other higher antibiotics may be tried. It is better to eradicate the disease by testing the birds and sacrificing the positives.

9.2. FOWL TYPHOID

Etiology: The disease is caused by *Salmonella gallinarum*. In many respects the organism acts like that of *Salmonella pullorum*. Most species of poultry susceptible to pullorum are also susceptible to typhoid. The disease may affect birds of any age but it is more commonly seen in the birds of about 3 to 6 months.

Host: Turkeys, guinea fowls, quails, pheasants and wild birds are also known to suffer from the disease. The cycle of infection is similar to that of pullorum disease. *S. gallinarum* may be found in 10 to 15 percent eggs laid by a carrier hen. Feed and water play an important role in spread of infection.

Symptoms: Affected birds may show dullness, loss of appetite, fever and yellowish diarrhoea. Death of birds takes place from 2nd day and declines by about 5th day and usually stops by about 8th month. The wattles and combs become cyanotic. The disease may reoccur subsequently but with reduced virulence and mortality. The disease in chicks may show signs similar to those of pullorum disease.

PM Lesions: In peracute cases, no appreciable lesions may be observed, but in acute cases the spleen and liver get enlarged with a characteristic copper colour of liver. As in the case of pullorum disease, necrotic spots may be seen in fowl typhoid in the liver, lungs and gizzard. Catarrhal enteritis with or without ulcers may be also found. The ova may get deformed and discoloured. In chronic cases sinovites and arthritis may also occur. Heart may show small grey nodular granulomas. Lungs may show greyish areas of consolidation and heart may reveal necrotic grey spots.

Diagnosis: The diagnosis is the same as for pullorum. Agglutination tests may be used to determine if adult birds are carriers. A laboratory diagnosis is the only accurate means of determining the presence or absence of *S. gallinarum* organisms.

Prevention and Control: Eradication of the carrier birds in the breeder flock should be the first line of attack in handling an outbreak of typhoid in young chicks. Depopulation of the entire breeder house is not a practical means of control. For disinfection 1:1,000 phenol; 1:20,000 mercuric chloride or 1 per cent potassium permanganate may be used. Fumigation may also be done after infection of premises.

Treatment: As with pullorum, furazolidone is the drug most often used in treating the birds affected with fowl typhoid. Antibiotics seem ineffective. Do not feed furazolidone to the replacement pullets older than 14 weeks of age or to laying hens.

9.3. PARATYPHOID

Etiology: *Salmonella enteritidis*; *S. typhimurium* and *S. montevideo* have been traced to affect the chicken causing serious losses. The paratyphoid infections can occur in most species of warm and cold blooded animals and among poultry the most common hosts are turkeys and chicken.

Symptoms:

i) **Young Birds:**—Paratyphoid infection is most common among young birds. In acute form, no signs may be exhibited except high mortality of embryos in hatcher or after few days of hatching. In less severe form, the symptoms include somnolence, birds standing with lowered head, droopy wings and closed eyes, ruffled feathers, anorexia, increased water intake accompanied with watery diarrhoea and vent pasting, huddling near the heat source and rarely respiratory signs. In some cases blindness may result due to vascuolation of corneal membrane and corneal opacity.

ii) **Adults:** Usually no gross symptoms are seen and they act as potential carriers. The mortality also seldom exceeds 10 percent.

Transmission: In generally, transmission is similar to that of *S. Pullorum* but certain modes are of more importance.

i. Egg Transmission: Paratyphoid definitely is egg transmitted but egg shell penetration is more important than in case of *S. pullorum* and *S. gallinarum*. Egg shells are abundantly covered with paratyphoid organisms as the egg passes through the cloaca, when the egg is laid these organisms are sucked through the shell pores, an avenue of embryo inoculation.

ii. Faecal Contamination: The paratyphoid organisms are found in large numbers in intestinal tract which accounts for greatest amount of transmission from bird to bird.

iii. Ovarian Transmission: Paratyphoid organisms lodge in the ovary; this represents a possible method of transmission of the disease from dam to the newly hatched chicks. In most cases however, there is little ovarian transmission. Young chicks can be infected but will not show evidence of the disease until they are stressed.

iv. Feed: As with pullorum disease, the paratyphoid organisms may live in certain feed ingredients.

PM Lesions:

i) **Young Birds:**—In very severe outbreaks, there may not be any lesions. In less severe cases the common lesions are emaciation, dehydration, coagulated yolks, congested liver and spleen with haemorrhagic streaks or pinpoint necrotic foci, congested kidneys and pericarditis with adhesions. Heart and lung lesions are not common.

ii) **Adults:**—In acute infection, the lesions that may be seen are congested and swollen liver, spleen and kidneys and haemorrhagic or necrotic enteritis, pericarditis and peritonitis. In some cases, ovary and oviduct may also be involved.

Diagnosis: Gross symptoms and lesions with a supportive history may help drawing a tentative diagnosis. Isolation and identification of organisms confirms the disease.

Treatment: Many paratyphoid carriers give positive reaction during salmonella blood testing and therefore, most of the carriers would have been automatically eliminated. Besides treated birds usually act as carriers since the therapeutic measures are not capable of eliminating the infection.

Prevention and Control: Infected and recovered flocks or carrier birds should not be used for breeding. Sanitation of poultry premises, hatching eggs and feed are also extremely important. Serological testing by macroscopic tube agglutination test or by rapid serum plate test or by rapid blood test may be practised.

9.4. COLIBACILLOSIS:

The coli organisms are responsible for a variety of poultry diseases with a variation in manifestation. The Escherichia coli are bacteria that represent one of many of the coliform group of organism's inhabitating the lower intestinal tract. Most are harmless and are called saprophytic and others are pathogenic and produce certain poultry diseases namely:-

Air-sac Infection
Coli Enterites
Coli Granuloma
Coli Septicemia
Egg Peritonites
Synovites
Yolk Sac Infection.

Coli Enterites: The organisms located in the upper portion of the intestinal tract cause it to become congested with small blood vessels. Deadly toxins are produced. Vessels rupture, causing haemorrhages similar to coccidiosis. Nodules also may appear in the caecal lining but whether the coli organisms are primarily responsible for the intestinal and caecal disorders is not known.

Coli Septicemia: When toxins and bacteria enter the blood stream after the toxin produced by the coli organisms rupture intestine wall. Such lesions allow the organisms to gain entry in the portal system and finally enter kidneys which become congested and enlarged. The liver is next to be affected which becomes enlarged showing its discoloration.

Air-sac Infection: Eventually E. coli involve the air sacs, the organisms arriving by the way of blood stream. Air sacculitis results and the birds cough and wheeze. Morbidity rather than mortality becomes the economic problem especially in broilers; the air sacs become filled with yellowish cheesy material which can be also observed around heart and lungs.

PM Lesions: The most important lesion is the presence of milky fluid in the pericardium due to pericarditis. The air sac membranes may become thicker and cloudy in appearance. Liver may show a thin covering of fibrinous exudate and liver becomes dark.

Transmission: There are several means of E. coli transmission.

Faecal: Organisms in the intestinal tract are continuously being sluffed through the faecal material and in turn these bacteria dry and float in the air and gain entry to uninfected individuals through respiratory tract and intestinal tract.

Egg Shell Contamination: As the completed egg lies in the cloaca prior to being laid, it becomes contaminated with the excrement of the intestinal tract including E. coli. Subsequently, some organisms enter the egg contents and reach the developing embryo thereby affecting the hatchability and chick quality.

Respiratory: As air sac infection from E. coli can be the result of infection through the respiratory tract contaminated dust in the poultry house can be a direct cause of transmission.

Ovarian: Transmission through the ovary is possible when birds are shedding the *E. coli* organisms through uterine infection. Infected breeder hens thus transmit the disease to the newly hatched chick.

Feed: Although not a primary route of infection coliforms may gain entrance to the body through contaminated feed.

Diagnosis: A laboratory test is the only satisfactory method of accurate diagnosis. Coliforms are isolated and classified.

Treatment: Any treatment must begin with a cleanup campaign as most *E. coli* infections start with dirty surroundings.

Sulphadimethoxine plus ormethoprim is the only feed treatment recommended for colibacillosis. Other drugs have been used including tetracyclines, sulfas, novopiocin and gentamicin. An antibiotic sensitivity test may be helpful in determining effective medications.

9.5. OMPHALITIS (NAVAL INFECTION):

Omphalitis is a disease of general bacterial infection due to several organisms. They may be coliforms, staphylococcus, pseudomonas or other types. Bacteria invade the umbilical tissues as the result of improper conditions in the hatcher. The naval opening does not close and infection passes to the internal organs.

Symptoms: The chicks seem weak, huddle together and may have watery diarrhoea. An infected and open umbilical area will be noticed. It will be discolored to bluish black. There is a very noticeable pungent odour characteristic only of this disease. The abdomen feels soft, mushy, flabby and enlarged. The infection may carry to the internal organs, particularly to a portion of the intestines. Peritonitis may be found when the yolk sac ruptures. Mortality may run as high as 10%.

Transmission: The disease is very infectious and death occurs within 2 or 3 days after hatching. The seat of the infection is the incubator (hatcher) although the original source probably is bacterial contamination of the egg shells before the hatching eggs enter the hatchery. The ventilating fans quickly disseminate the organisms which find the unhealed naval a likely seat for infection.

Diagnosis: Suspected chicks should be submitted to the laboratory for diagnosis. A bacteriological examination of the yolk sac contents will identify the disease and the causative organisms.

Treatment: If the chicks in the brooder house appear chilled, increase the brooding temperature. Although antibiotics or nitrofurans may check the disease in the brooder house, the possibility is very remote.

Control: When an outbreak occurs in the hatchery, equipment and rooms must be fumigated with formaldehyde gas. Use 3X strength where possible. Lower concentrations are not effective in destroying all responsible organisms. Incubating eggs (1 to 19 days) should be fumigated at 2X (double) strength never, more for this procedure.

Hatchery rooms and all equipment must be fumigated every second day until the infected is destroyed. Also use a good liquid disinfectant where practical.

9.6. MYCOPLASMA GALLISEPTICUM:

This disease is known everywhere and is extremely important to both the broiler grower and layers. Infection of the air sacs in the broilers is cause for condemning the dressed birds as unsuitable for human consumption.

Etiology: *Mycoplasma gallisepticum* (MG) organism is very small and delicate and has no cell wall. MG is a stress disease because the organism seems to remain dormant in many flocks but when birds are stressed it becomes active. More MG develops in cool or cold environment.

Symptoms:

Young Birds: MG is a respiratory disease affecting the entire respiratory tract particularly air sacs where it localizes. All the air sacs may become involved are cloudy in appearance and filled with mucus. In the later stages of disease, this mucus develops a yellow colour and a cheesy consistency. Similar exudates may encircle the heart. In young chicks there is rattling, sneezing and sniffling, all indicative of the respiratory difficulty.

Adult Birds: Visible evidence of the disease in adult birds may go unnoticed. Occasionally the birds will appear depressed and inactive. There may be a definite diarrhoea during the intestinal phase of the disease. Egg production may suffer.

PM Lesions: Predominant lesions are as follows:-

- Mucoid tracheitis.
- Fibrinopurulent pericarditis.
- Thick fibrinopurulent exudates on liver.
- Air sacculitis.
- Purulent synovitis of hock joints.
- Congestion and dark foci in lungs.

Diagnosis: MG may be fairly accurately diagnosed by the coughing and sniffling. Laboratory tests used for diagnostic determination are:-

- Rapid serum plate or tube agglutination test.
- Haemagglutination inhibition test.
- Embryonic examination.

Treatment: Tylosin is an antibiotic specific for the treatment of the disease. Auremycin, erythromycin, spectinomycin, doxycycline or streptomycine.

9.7. ARIZONOSIS:

Arizonosis is an important disease of turkeys and less important for poultry.

Etiology: The disease is caused by gram negative, motile bacteria called *Salmonella arizonae.*

Symptoms: The disease affects chicks below 3 weeks of age and peak mortality occurs between 5th and 10th day of age. Adult birds do not show any symptoms but young chicks show diarrhoea, vent pasting and paralytic signs due to dehydration.

PM Lesion:

- Duodenum is inflamed.
- Caeca may contain cheesy plugs.
- Yolk may be unabsorbed.
- Liver is enlarged, yellowish or mottled.
- Small abscesses may be found in lungs.
- There may be cloudiness of cornea or a thin layer of cheesy material may be seen on the corneal surface.

Prevention and Treatment: It is done on the same lines as for paratyphoid. Furazolidone may be given at the rate of 100gm per tonne of feed for 1 week after hatching to prevent the disease. For treatment the dose may be doubled. Other drugs suggested for treatment of paratyphoid can also be used for this disease.

9.8. ASPERGILLOSIS (BROODER PNEUMONIA):

Etiology: This disease mainly affects the respiratory system but sometimes infection may spread to the other visceral organs. *Aspergillus fumigatus* is the most common etiological agent. Other species like *A. flavus, A. nidulans, A. glaucus, A. niger* and *A. candidus* have also been found related to the disease. Normally this fungus grows on decaying organic matter in the poultry farms and in the hatchery but it also has the ability to reproduce itself in certain tissues of the bird.

Symptoms: The lungs are the major areas of internal infection. A close examination of the lung tissues will show small nodules that are hard and yellow. There may be but a few in some cases, in others there may be hundreds. In some instances the fungus gain entrance to the air sacs and a respiratory condition evolves. However, there is no coughing or sneezing.

Chicks have few external symptoms, except for an occasional involvement of the eye causing semi-blindness. As the fungus continues to grow in the lungs and air sacs, flock mortality increases. It may be quite high in young chicks. Older birds are able to withstand the fungal growth and few birds are affected.

Transmission: Spores from the fungus dry and are transported easily from chick to chick by way of the air. This may occur in the hatchery or in the brooder rooms. The incubator may become a source of contamination.

PM Lesions: Lesions are well developed in the birds above 5 days of age. Lungs show almost uniform raised pin head sized yellowish nodules. Air sacs become thickened and cloudy, showing yellowish plaques. Necrotic foci may occasionally be seen in the liver, spleen, kidneys, proventriculus and other organs sometimes the fungus may invade the walls of the blood vessels, brain, liver, heart, respiratory tract, spleen, proventriculus and kidneys showing granulomatous lesions.

Diagnosis:

- By characteristic gross lesions.
- Demonstration of fungus by lactophenol cotton blue stain. The caseous material of the nodules in the lungs or air sacs mixed with this stain shows bluish segmented filaments of fungus under the microscope.
- Fungus can be cultured on saboraud's agar or tryptiease soy blood agar at 35°C.

Prevention: Feed with low moisture content and provide dry litter in the shed. If the disease has been developed, then the infected litter and feed should be disposed off or preferably burnt and disinfection of brooder may be done with caustic soda solution.

Treatment:

- Administer tetracycline sorbate at 200 mg per lit water for 5 days.
- Use hamycin at the rate of 10 ml suspension per liter of drinking water for 10 days.

9.9 NEWCASTLE DISEASE:

This is the most dreaded disease of poultry and causes very heavy mortality at a very fast speed. In India it is known as Ranikhet disease.

Cause: The disease is caused by a virus of avian paramyxotype I. There are four general forms classified according to their pathogenicity. The virus may be neurotrophic (nervous) pnemotropic (respiratory) or viscerotropic (internal organs). Death in the sub acute form is usually the result of paralysis. In some cases the organisms invade the respiratory tract; in others, they may locate in the intestines and proventriculus.

Host: Newcastle disease occurs in most virulent form in poultry and less acute in ducks, turkey and pheasants. Some wild birds like pigeons, doves, crows, sparrows, parrots, kites and vultures can suffer and spread the disease to poultry farms.

Symptoms: Symptoms vary according to the age of the bird and the form of Newcastle disease virus involved. Three major symptoms usually are:-

- Respiratory signs.
- Nervous signs.
- Egg production and egg shell quality is reduced.

The virus localizes in the respiratory tract and all affected birds show evidence of the respiratory type. If nervous symptoms are involved, they arise later. Older birds seldom show any manifestation of nervous disorders. Egg production and egg shell quality are affected quickly in laying birds. There are four forms of the disease and the symptoms for each are:

- Viscerotropic velogenic (VVND) (Exotic ND): Very highly pathogenic. It is sometimes known as the Asiatic type. It is highly virulent with high mortality. Respiratory and nervous signs are less evident. Young chicks show spasms and twisted necks.

- Velogenic or neurotropic (field type): It is highly pathogenic. The affected birds show sudden onset, it is acute and often fatal. Birds show high morbidity. Evidence of nervousness (twisted neck). There is difficulty in respiration.
- Mesogenic (Intermediate pathogenicity): The disease shows acute signs in young chicks but not in older birds.
- Lentogenic (mild pathogenicity): All ages of birds may have unnoticed infection with existence of mild respiratory signs. Egg production gets declined with rapid deterioration in egg shell quality. In acute or velogenic forms mortality may reach up to 90 to 100% but if temperature of the environment is low it may go up to 55%. Milder forms may show mortality from 5 to 50%.

PM Lesions: In Doyle's form the most common lesions are ulcers in the intestines which are elongated antero-posteriorly and can sometimes seen even without opening the intestines. There are haemorrhagic ulcers almost invariably found in caecal tonsils. The proventriculus may show pin point haemorrhages on the serous membranes and heart. The ulcers of intestines are haemorrhagic, raised and generally with thick necrotic surface. In mesogenic form of the disease spleen becomes small and mottled. In lentogenic form of the disease, the haemorrhagic lesions are few or absent.

Transmission: New castle disease virus is spread by the following means:-

1. Through the air: Coughing dislodges the virus from respiratory tract whence it is becomes an air borne infection.
2. On clothing, unsanitized equipment: This category probably represents the major means of transfer of the virus from infected flocks to uninfected flocks and forms.
3. No cleanup period on farm: Poultry operations that stocks chicks on a regular basis results in several ages of birds on the farm at one time. Farms with an all-in-all-out programme of management break any cycle of infection.
4. Feed.
5. Wild birds.
6. Exotic birds.
7. Predators.

Diagnosis: Many times a diagnosis may be made on the basis of physical observation. Nervous disorders inflicting twisted necks and causing the chicks to roll or tumble are enough to warrant a fairly accurate diagnosis. When the diagnosis becomes more difficult certain laboratory techniques must be employed.

1. Haemagglutination test.
2. Virus isolation test.
3. Haemagglutination-inhibition test.
4. Fluorescence antibody test.
5. Enzyme linked immunosorbent assay test.

Treatment: There is no known treatment, although broad-spectrum antibiotic medication for secondary disease will probably reduce some of the flock morbidity.

Prevention and control: Three types of vaccines are available in India:

* F1 vaccine—One drop I/O or I/N given at 4-5 days of age.
* R2B vaccine—0.5 ml I/M given at 15-16 weeks of age.
* Lasota—given at 3 to 4 weeks of age through drinking water.

9.10 INFECTIOUS LARYNGOTRACHEITIS (ILT):

Etiology: Infectious laryngotracheitis is an acute or sometimes sub acute disease of respiratory system which is caused by a DNA herpes virus. Symptomless carrier birds are known to carry the virus for as long as 16 months which may be excreted through their faeces.

Host: ILT is mainly a disease of poultry affecting the adult birds but mild in chicks below 3 weeks of age.

Transmission: The disease spreads rapidly from bird to bird by following modes:-

- By air: Although airborne for short distances as in a pen of birds, the organism cannot be carried in the air over long distances.

- By people, trucks, birds, rodents: Mechanical means of dissemination of the virus is no doubt, the major means of spread.

Symptoms: Incubation period of the disease is about 6 days. The birds below 3 weeks of age show only conjunctivitis or watery fluid between the eyelids. In the adult, the acute form of disease is accompanied by difficult respiration. During inhalation of air the birds lower their head and exhale. They extend the neck upwards with beak wide open. The wattles may become bluish. There may also be sneezing, coughing and rales. Blood tinged exudate may come out during coughing. In subacute form of disease the sinus and discharge of exudate from nostrils. The mortality in mild form of disease in adults remains around 10 to 20% but acute form it may reach up to 70%.

PM Lesions: A cheesy material may be found in the nasal cavity, intra orbital sinus and slight congestion of larynx and trachea in sub acute form in adults. In acute form, the larynx and trachea shows acute inflammation and blood tinged exudate in them. Sometimes the mucous may be necrosed and cheesy exudate may be found in them.

Diagnosis:

1. The symptoms give some indication of the disease.
2. The characteristic lesion of larynx and trachea give further support to the diagnosis.
3. Virus isolation.
4. Agar gel precipitation test.
5. Fluorescent antibody technique.
6. Neutralization test.

4Treatment and Control:

1. If an outbreak occurs, then slaughter and proper disposal of birds is the only method.
2. Terramycin, auromycin etc. may be used to prevent secondary infections in the outbreak of disease.
3. Disinfection of infected premises may be carried out with 1% caustic soda or 3% cresol.

9.11 AVIAN LEUKOSIS COMPLEX:

Avian leukosis (AL) is a worldwide disease of cancerous type which may produce tumors or disease of blood forming cells, lymphoid tissue, bone, kidneys and connective tissue.

Etiology: The virus causing AL is a, C-type RNA retro virus of subfamily Oncovirinae. The ALC virus of chicken are divided into 5 subgroups A, B, C, D and E depending on their antigenic differences. The A and B subgroups of virus are important in causing disease in poultry under field conditions. A and E subgroups of viruses are found in most of poultry farms and in some farms A and B subgroups are found together.

Hosts: The disease is mostly seen in poultry, affecting birds which have attained sexual maturity as compared to MD which occurs often before sexual maturity.

PM Lesions: ALC can take one of the several pathological types of the disease but lymphoid leukosis represents tumours of lymphoblastic cells is the most common.

Lymphoid leukosis: In this form, the virus first multiplies in the bursa of fabricious causing tumors of bursa, 5-6 weeks after infection. These tumors are most commonly found in liver and spleen but may also be seen in lungs, ovary, serosal surface. The tumor cells may cause diffuse enlargement of liver and spleen.

Erythroblastosis: Sometimes the ALC virus may affect the erythroblasts of bone marrow causing flooding of the bone marrow with tumorous erythroblasts.

Myeloblastosis: This form of disease may occasionally affect younger birds. In this form there is neoplastic multiplication of granular leucocytes which spread to other internal organs causing enlargement of liver, spleen and kidneys.

Myelocytomatosis: This form of disease generally occurs between 3 to 11 weeks of age. In this form, more mature granular leukocytes become tumorous affecting the internal surface of the ribs, sternum or pelvis and sometimes in liver, spleen or kidneys.

Transmission: through the eggs.

Diagnosis: The birds affected with ALC hardly show any characteristic symptoms except in osteopetrosis form. The diagnosis is mainly on tumorous enlargement of liver, spleen, kidneys, ovaries etc. which may be diffuse or nodular. There are large numbers of tests which can be used for the laboratory diagnosis as:

1. Resistance inducing factor test.
2. Complement fixation test.
3. Non producer test.
4. Fluorescent antibody test.
5. Neutralization test.
6. Precipitation test.

Prevention and Control: ALC is not curable. The virus is present in almost every flock and is difficult to keep the flock ALC virus free because of egg transmission. No vaccine so far is available against the disease.

9.12 AVIAN INFECTIOUS BRONCHITIS:

Infectious bronchitis (IB) is a highly infectious viral disease of poultry which causes respiratory signs or drop in egg production or decline in body weight.

Etiology: The virus causing IB is a corona virus of about 100 nm size. It is a RNA virus. IB may be complicated by *M. gallisepticum*, *E. coli*, Adeno or Reovirus infection.

Host: IB affects only poultry and can produce disease in birds of any age but the baby chicks are more susceptible to the respiratory form. Heavy breeds are more susceptible to nephrotic form while lighter breeds to the respiratory form.

Symptoms: Small chicks show signs of sneezing, coughing, respiratory rales or even dyspnoea. Some chicks may show swelling of head. Eyes may look wet. Birds may become dull and tend to huddle towards the source of heat. Diarrhoea may be seen after a few days. Mortality is maximum within the first and second week after infection and tapers off by about the end of the third week. Adult birds show sudden drop in egg production to as low as 25 to 50 per cent. Egg production may be delayed by 4 weeks. Egg production may return to normal after 1 to 1½ months. The nephrotic form of the disease affects birds under 2½ to 3 months of age. Heavy breeds are more susceptible to nephrotic form while light breeds to the respiratory form. The quality of eggs may not be normal and hatchability may also decline. Some of the eggs may be thin shelled, rough shelled, deformed or smaller than normal. Mild sneezing, rales and coughing may be seen in some layers.

Transmission:

1. Air borne.
2. By people, birds and animals.
3. By equipment.
4. Through feed.
5. Through carrier birds.

Diagnosis: Diagnosis of infectious bronchitis is difficult. It is often made by eliminating the incidence of other similar diseases as causative agents. The following tests are employed to diagnose the disease.

- Serum neutralization test.
- Virus isolation test.
- Haemagglutination test.
- Fluorescent antibody test.
- Enzyme linked immunosorbent assay test.

Treatment: There is no known treatment for infectious bronchitis. However, when secondary infections are evident. Treatment for these may alleviate the damage to the bird.

Prevention and control: Vaccines against IB are produced in India by some private manufacturing companies. Modified Massachusetts strain can be installed as a drop into the nostrils at 5-10 days of age. Booster dose is given in the same way in growers and broilers through drinking water at 3-4 weeks of age. 2nd vaccine as booster dose is given by 13-15 weeks of age.

9.13 AVIAN ENCEPHALOMILITIS (AE):

Avian encephalomilitis also called epidemic tremors is caused by Picorna virus. It is a RNA virus. The disease affects mainly poultry and experimentally turkey, pheasants, ducklings.

Symptoms: The new born chicks hatched from infected eggs show paralysis in the first week after hatching. Tremors of muscles especially of head and neck may be seen in diseased chicks. Tremors can be aggravated by shaking the birds or handling them. Some birds may show blindness due to opacity of lens which may persist up to 6 months. Death may occur in one day or several weeks after the onset of symptoms. Older birds show only drop in egg production. AE viruses have also been known to cause significant cases of enteric disease.

Transmission: AE is an important egg borne disease.

PM lesions: No gross lesions are observed in the chicks dying due to AE except the presence of minute grayish spots in the muscles of gizzard or proventriculus.

Diagnosis: Age of the affected birds, symptoms and histo-pathological lesions of brain, spinal cord and lymphoid aggregates especially in muscles of heart, proventriculus, gizzard and pancreas are helpful in establishing tentative diagnosis.

Prevention and control: There is no treatment for AE. The recovered birds show poor growth and production.

9.14 EGG DROP SYNDROME-1976

Etiology: Adenovirus (DNA) exists as only one serotype usually affects the chicken or goose adenovirus and that chicken is not a natural host. Chicks can be experimentally infected by ducks only when they are kept in close contact which under commercial conditions is very rare.

Transmission: Main route of transmission is through embryonated egg. Horizontal transmission is slow and intermittent especially in caged birds. Spread appears to depend on birds coming into contact with infected faeces.

Symptoms: When the virus has been transmitted horizontally, the first sign is loss of colour in pigmented eggs followed quickly by production of thin shelled eggs, soft shelled or shell less egg. The thin shelled eggs often have a rough sand paper like texture or a granular roughing of the shell at one end. There will be a sudden drop in egg production if infection has occurred in late production. If affected eggs are removed, incubation performance is unaffected. If the disease is due to reactivation of virus, small eggs and eggs with poor interior quality are produced. If birds have acquired antibodies before the latent virus is unmasked, then the birds do not attain the expected peak in egg production and there may be delayed sexual maturity. Rate of neither morbidity nor mortality has not been accurately defined.

PM Lesions: In natural outbreaks, the only findings in some birds may be inactive ovaries and atrophied oviducts.

Diagnosis: Isolation and identification of the virus and serological tests.

Prevention and Control: Since the disease is egg borne, birds must be obtained from uninfected flocks. Ducks should not be reared in close proximity with chickens. For immunization an oil adjuvant inactivated vaccine is widely used and the birds are vaccinated between 14 and 16 weeks of age.

10

Hatchability Troubleshooting

Any investigation of the causes of poor hatchability must include examination of dead in shell. Several poultry diseases that affect the parent breeder flock have an effect on the developing embryo, hatchability and chick quality. Therefore, it is almost impossible to differentiate the source of infection by observation of dead embryos or the newly hatched chicks. Thus, the main points to look for are:

- Egg size and shell quality.
- Air space.
- Position of embryo within shell.
- Anatomical abnormalities.
- Nutritional abnormalities.
- Unused albumen.
- Age of albumen.

Major causes of eggs failing to hatch are as follows:-

- Egg storage.
- Breeder nutrition.
- True infertility (flock age).
- Disease.
- Bacterial and mould contamination.
- Genetics.
- Egg faults and shell damage.
- Incubation faults.

Investigating problems of poor hatchability is often confounded by the fact contributing events may have occurred weeks or even months previously. Unless the hatchery is undertaking weekly breakout of fresh eggs, then it is often hatch results from eggs laid 3-4 week previously that are suspect. The main question to answer relates to egg being fertile or not and while this sounds like a relatively simple question, there is often confusion during diagnosis at the hatchery. Once an egg has been incubated for just a few days, it becomes increasingly more difficult to differentiate between an infertile and an early dead germ. In the warm moist environment of the incubator there is a rapid change in yolk colour, consistency and appearance and so identification becomes more difficult. Fertility is best assessed at the farm by breaking out fresh eggs. In this situation, the fertile egg is characterized by a raised 'doughnut' shaped ring of lighter colour than the surrounding yolk. If fertility is less than standard, then the check list (Table 10.2) should be followed. Fertility problems usually relate to the male, since if the hen lays an egg then there is little chance óf it being infertile if sperms are present in the oviduct. However if the hen is in generally poor condition, the mating activity may be reduced and the patterns of ovulation and fertility may be adversely affected. High temperatures are a fairly common cause of infertility. The male is particularly affected by heat stress and infertility may not occur until 10-14 day following the high temperatures, and low sperm production may continue for 3-4 weeks following the actual stress conditions. Maintaining effective (actively mating) males is much less than 6 per 100 females. This problem caused by high male mortality can only practically be resolved by introducing new males into the flock. If reduced hatchability is due to early dead embryos (less than 5 day incubation) then such problems most likely relate to on farm conditions. It is unusual for the very young embryo to die because of improper incubation conditions, unless there are extremes in temperature. The very young embryo is however quite sensitive to pre-incubation temperature, humidity and physical handling. Pre-incubation can occur if eggs remain in the nest too long, especially in warm climates and this over developed embryo is most likely to die during subsequent cooling and reheating during actual incubation. Pre-incubation most commonly occurs if eggs are not cooled adequately at the farm or during transport to the hatchery.

Once eggs have been sorted and cleaned, they should be ideally cooled to 19-20°C as soon as possible. Such cooling can take 2-3 days to occur if eggs are boxed prior to storage at the farm and resultant pre-incubation can lead to embryo mortality. Unfortunately many breeder farms are located some distance from main highways (for biosecurity) and this means that egg trucks have to travel considerable distances on farm roads. Again the very young embryo is susceptible to excessive physical movement and such movement of eggs does lead to some loss in hatchability.

It is unusual for breeder nutrition to be a factor in early embryo mortality. Unless a breeder diet is grossly deficient in any nutrient (in which case egg production will also be affected) then the embryo can grow for 7-10 days on very meagre reserves of most vitamins and minerals. Such deficiencies therefore, more commonly result in mid-dead embryo mortality, which occurs characteristically in the 10-14 day period of incubation. During the very early stages of nutrient deficiency embryos may infact develop almost completely and so the first signs of this are late (18-20 day) mortality. As the deficiency proceeds then mid-dead embryo mortality quickly becomes the characteristic feature.

Problems with late dead embryo mortality are usually associated with incubator conditions and are rarely farm related.

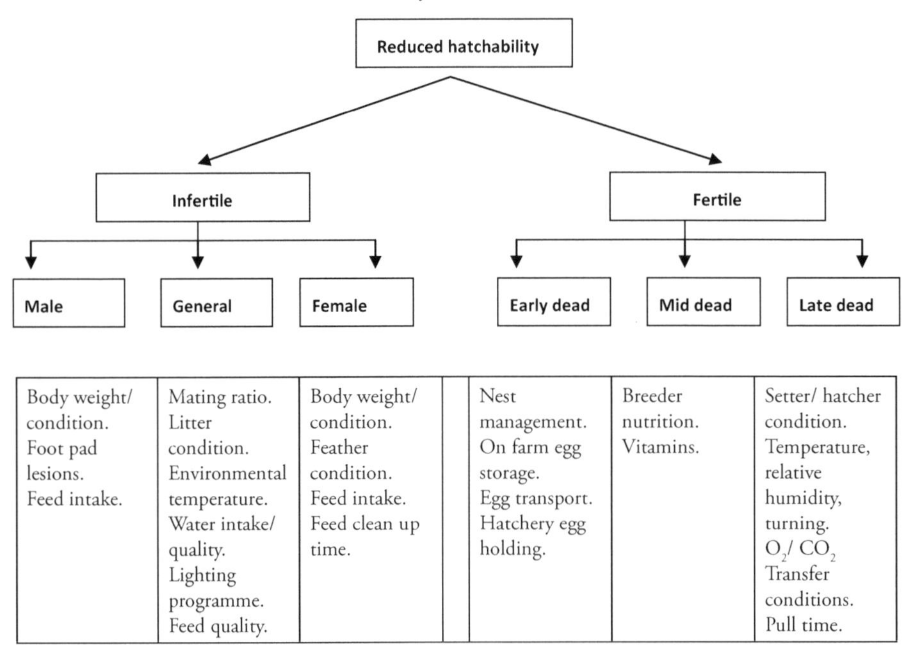

Male	General	Female		Early dead	Mid dead	Late dead
Body weight/ condition. Foot pad lesions. Feed intake.	Mating ratio. Litter condition. Environmental temperature. Water intake/ quality. Lighting programme. Feed quality.	Body weight/ condition. Feather condition. Feed intake. Feed clean up time.		Nest management. On farm egg storage. Egg transport. Hatchery egg holding.	Breeder nutrition. Vitamins.	Setter/ hatcher condition. Temperature, relative humidity, turning. O_2/ CO_2 Transfer conditions. Pull time.

Fig. 10.1: Probable causes of low or poor hatchability

EGG BREAKOUT:

Whatever system of breeding and male/female management is used, the success of the reproductive process is established by egg breakout at the farm and/or hatchery. Egg breakout is used to assess fertility prior to incubation and also early dead germs during the first 5-7 day of incubation.

For fresh eggs the shell is carefully broken and yolk rolled within the shell or on the palm of the hand until the germinal disc is clearly present. The fertile egg has a clearly visible raised disc (doughnut) which is white or of less intense colour than the surrounding yolk. For the unfertilized egg, the germinal disc is evident but lacks texture. True fertility is calculated as the number of fertile eggs observed expressed as a percentage of all eggs examined. For any breeder flock, at least 60 eggs should be examined.

Eggs can be examined after 5-10 day incubation to determine fertile/ infertile numbers as well as occurrence of early dead germs (Table 10.1). Eggs are carefully broken and the germinal disc/embryo examined within the broken shell or again on the palm of the hand. Infertile eggs retain the original yolk colour and are little different in appearance compared to fresh eggs. The germinal disc is still quite distinct and small. With embryos dying very young, the germinal disc will be larger in diameter and there will not be a distinct raised disc per se. The yolk is often paler in colour and sometimes has a mottled appearance. When candling at 7 day, the total number of viable embryos, expressed as a percentage of all eggs examined is often referred to as candling fertility. Eggs can be examined after chicks have been taken from the hatch trays to determine general age of embryo mortality. The most common classification is as early, mid and late dead germs which refer to 1-7, 8-14 and 15-21 day respectively.

It is more difficult to differentiate infertile from very early dead embryos after the eggs have been incubated for 21 days. Determining age of embryo mortality is important for problem solving and identifying. For example any contribution of poor breeder management or nutrition. As a generalization, assuming that no disease situation is involved then early dead germs are related to important pre-incubation egg handling and/or egg storage conditions. On the other hand, mid dead germs (7-14 day) are rarely seen under normal conditions, although we can create a very high incidence by feeding diets inadequate in certain vitamins.

Mid-dead germs are therefore, a clue to inadequate diet formulation or vitamin antagonists being problematic. Late dead germs are more likely caused by incubation conditions and rarely relate to breeder management or nutrition.

TABLE 10.1: PATTERN OF EMBRYO MORTALITY RELATED TO BREEDER AGE

Description	Days of incubation	Embryo identification	Expected occurrence (%)		
			27 wks	45 wks	64 wks
Infertile	0	Germinal disc	10	7	15
	1	Primitive streak			
Early dead (1-7 d)	3	Embryo on left side	4	2	4
	5	Appearance elbow/knee joint			
	7	Comb evident			
Mid dead (8-14 d)	11	Tail feathers			
	13	Feather down	0.7	0.5	0.5
	16	Feathers			
Late dead (15-21d)	18	Head under right wing			
	20	Yolk sac absorbed			

(Source: Chick Master, P.O. Box 704, Medicine, Ohio 44258, USA)

TABLE 10.2: TROUBLE SHOOTING CHART FOR INCUBATION OF EGGS.

Observation	Possible Causes
Eggs exploding	Bacterial contamination of eggs. Dirty eggs. Improperly washed eggs. Incubator infection.
Clean eggs	Infertile. Eggs held improperly. Too much egg fumigation. Very early embryonic mortality.
Blood ring (Embryonic death, 2-4 days)	Heredity. Diseased breeding flock. Old eggs. Rough handling of hatching eggs. Incubating temp too high. Incubating temp too low.
Dead embryos (2nd week of incubation)	Inadequate breeder ration. Disease in breeder flock. Eggs not cooled prior to incubation. Temperature too high in incubator. Temperature too low in incubator. Electric power failure. Eggs not turned. Too much CO2 in air.
Air cell too small	Inadequate breeder ration. Large eggs. Humidity too high (1-19 days).
Air cell too large	Small eggs. Humidity too low (1-19 days).
Chicks hatch early	Small eggs. Leghorn eggs vs meat-type eggs. Incorrect thermometer. Temperature too high (1-19 days). Humidity too low (1-19 days).

Chicks hatch late	Variable room temperature. Large eggs. Old eggs. Incorrect thermometer. Temperature too low (1-19 days). Humidity too high (1-19 days). Temperature too low in hatcher.
Fully developed embryo with beak not in air cell	Inadequate breeder ration. Temperature too high (1-10 days). Humidity too high (19th day).
Fully developed embryo with beak in air cell	Inadequate breeder ration. Incubator air circulation poor. Temperature too high (20-21 days). Humidity too high (20-21 days).
Chicks pipping early	Temperature too high (1-19 days). Humidity too low (1-19 days).
Chicks dead after pipping shell	Inadequate breeder ration. Lethal genes. Disease in breeder flock. Eggs incubated small end up. Thin shelled eggs. Eggs not turned first two weeks. Eggs transferred too late. Inadequate air circulation (20-21 days). Incorrect temperature (1-19 days). Temperature too high (20-21 days). Humidity too low (20-21 days).
Malpositions	Inadequate breeder ration. Eggs set small end up. Odd shaped eggs set. Inadequate turning.
Sticky chicks (albumen sticking to chicks)	Eggs transferred too late. Temperature too high (20-21 days). Humidity too low (20-21 days). Down collectors not adequate.

Sticky chicks (albumen sticking to down)	Old eggs. Air speed too slow (20-21 days). Inadequate air in incubator. Temperature too high (20-21 days). Humidity too high (20-21 days). Down collectors not adequate.
Chick too small	Eggs produced in hot weather. Small eggs. Thin, porous egg shells. Humidity too low (1-19 days).
Chicks too large	Large eggs. Humidity too high (1-19 days).
Trays not uniform in hatch or chick quality	Eggs from different breeds. Eggs of different sizes. Eggs of different ages when set. Disease or stress in some breeder flocks. Inadequate incubator air circulation.
Soft chicks	Unsanitary incubator conditions. Temperature too low (1-19 days). Humidity too high (20-21 days).
Chicks dehydrated	Eggs set too early. Humidity too low (20-21 days). Chicks left in hatcher too long after hatching.
Mushy chicks	Unsanitary incubator condition.
Unhealed naval, dry	Inadequate breeder ration. Temperature too low (20-21 days). Wide temperature variation in incubator. Humidity too high (20-21 days). Humidity too low after hatching.
Unhealed naval, wet and odorous	Omphalitis. Unsanitary hatchery and incubators.
Chicks cannot stand	Breeder ration inadequate. Improper temperature (1-21 days). Humidity too high (1-19 days). Inadequate ventilation (1-21 days).

Crippled chicks	Inadequate breeder ration. Variation in temperature (1-21 days). Malpositions.
Crooked toes	Inadequate breeder ration. Improper temperature (1-19 days).
Spraddle legs	Hatchery trays too smooth.
Short down	Inadequate breeder ration. High temperature (1-10 days).
Closed eyes	Temperature too high (20-21 days). Humidity too low (20-21 days). Loose down in hatcher. Down collectors not adequate.

(Source: Wilson, 1996)

11

Hatchery Sanitation

A sanitary hatchery is necessary for high hatchability and good quality chicks. The latter implies that chicks not only appear healthy at the time they are hatched but are free of many disease producing organisms that may attack the very young birds. Thus, hatching eggs must be produced by healthy breeding stock, the eggs incubated under sanitary conditions and delivered in clean vehicles.

DISINFECTANTS USED FOR HATCHERY OPERATION:

Surface disinfectants are most effective in the absence of organic material. Disinfection is not a substitute for cleanliness, it is a means of destroying microorganisms, but is effective only when things start out relatively clean.

CHARACTERISTICS OF DISINFECTANT:

All disinfectants should have following characteristics:

1. Highly germicidal.
2. Non-toxic to humans and animals.
3. Effective in the presence of moderate amounts of organic material.
4. Non-corroding and non-staining.
5. Soluble in water.
6. Capable of penetrating materials and crevices.
7. Should have no pungent odours.
8. Readily available and inexpensive.

CHEMICALS USED FOR DISINFECTION:

The chemicals used for surface disinfecting are many and their values are highly variable. Sanitizers may be grouped according to their base ingredients but many other factors affect their potency.

a) CRESOLS AND CRESYLIC ACID:

Cresols and cresylic acids are liquid yellow or brown coal tar derivatives. They have a strong odour, irritate the skin and turn milky when water is added but they have an excellent germicidal action. Many types are available. Some are used in conjunction with a detergent. They are effective against gram-positive and gram negative bacteria, most fungi and some viruses. They have greater use in the poultry sheds than in the hatchery. Cresol may be used for disinfecting floors or room surface areas, equipment and footbaths.

b) PHENOLS:

These are also coal tar derivatives with a base of carbolic acid and phenols have a characteristic odour, turn milky in the presence of water and are effective germicides. Most phenols are not compatible with the non-ionic (neither anionic nor cationic, but produce neutral colloidal properties) surfactants and should not be used with them. They are compatible with anionic (charged ionic compounds) and are more active in an alkaline pH because of their greater solubility. Their action is rapid.

These disinfectants are effective against fungi, gram positive and gram-negative bacteria but not effective against bacterial spores. They can act upon some viruses. They act as a protoplasmic poisons penetrating and disrupting the cell wall and precipitating the cell proteins when used in high concentrations, but only the essential enzyme system of the cell is disrupted at low concentration. Synthetic phenols may be used for egg dipping, hatchery equipment sanitation and footbaths.

c) IODINE:

Iodine compounds are available as iodophores which are combinations of elemental iodine and an organic solubilizing agent usually a non-ionic surface action agent that is soluble in water. The compounds react only with nucleic acid of the cell contents. They are good disinfectants in an acidic environment (2-4 pH) but activity diminishes in an alkaline pH and in the presence of organic material. They are effective against gram positive and gram negative bacteria attacking the nucleic acid of the organisms. Iodophores are also effective against fungi and some viruses.

d) CHLORINE:

Chlorine is an effective constituent of certain disinfectants. Sodium or calcium hypochlorites as powder form combined with hydrated trisodium phosphate are good disinfectants in which free chlorine is available to the extent of 200-300 ppm. Elemental chlorine or hypochlorites when added to water produces the bactericidal action but in presence of organic matter, chlorine combines with organic matter to form stable compound therapy reducing the free chlorine in solution. Chlorine is effective against bacteria and fungi and when coming from hypochlorites it attacks both the protein and nucleic acids of viruses. Bound chlorines as in chloramines have poor activity. Chlorine solutions are much more active in acid solutions than in alkaline and in a warm rather than in a cold mixture. Sodium hypochlorite is active but its disinfecting life is short. However, calcium hypochlorite is less active but its disinfecting quality for a long period. Chlorine compounds are somewhat irritating to the skin and corrosive to metals.

e) QUATERNARY AMMONIUM COMPOUNDS:

These compounds are cationic, odourless, clear, non-irritating, have a deodorizing and detergent action and are quite effective as surface disinfectants. The most common is alkyldimethyl benzyl ammonium chloride.

QACs are extremely water soluble but cannot be used in soapy solutions or where there is a residue of soap or anionic detergent. Their germicidal properties are reduced in the presence of organic material. These chemicals are effective against gram positive organisms, moderately effective against gram negative, effective against some fungi and viruses. Their effectiveness is increased by the addition of sodium carbonate.

A solution of 500ppm quaternary ammonium, 200ppm EDTA and sodium carbonate added to adjust the pH to about 8.0 is excellent disinfectant for hatchery. It may be used on floors, walls and incubator trays.

f) FORMALDEHYDE:

Formaldehyde reacts with both the non-nucleic acid protein coat and the nucleic components within the particle. A 10% formaline water solution is powerful disinfecting agent. Formaldehyde is primarily use as a fumigant in the hatchery.

Formaldehyde is commercially available as 40% solution in water as formaline and powder form as para-formaldehyde containing 91% formaldehyde which when heated liberates formaldehyde gas.

PROPORTIONS OF CHEMICALS:

The heat necessary for the release of formaldehyde from formaline is produced by mixing potassium permanganate with it. Two parts (by volume) of formaline are mixed with approximately one part (by weight) of potassium permanganate in an enamel ware or earthen ware vessel of large capacity because of boiling foaming splattering action when the two are mixed. Usually 40cc of 40% formaline and 20g of $KMnO4$ is sufficient for every 2.83 m^3 is known as 1X concentration. For the same area, if the quantities of the chemicals are doubled, it is called as 2X and so on. This will produce complete expulsion of the gas. When the reaction is complete, a dry brown powder will be left. If the residue is wet not enough potassium permanganate was used and if the residue is purple, too much permanganate was added.

Varying concentrations of formaldehyde gas are necessary to fumigate under different conditions. In the event that the fumigation must be stopped after a period of time, it can be affected, most of the times by opening the air intakes and exhausts. However it can be expedited by sprinkling the floor area fumigated with ammonium hydroxide. The quantity of ammonium hydroxide required is equal to one half of the quantity of formaline used for fumigation.

The efficacy of formaldehyde gas is increased in the presence of heat and moisture. A temperature of $24^{O}C$ or higher and relative humidity of 75% or more are ideal for achieving the best results. Varying concentrations of formaldehyde gas are needed under different conditions and the length of fumigation period is variable (Table 11.1).

The gas is detrimental to the living embryo and to the newly hatched chick. Therefore, care should be taken so that the concentration of gas and the length of fumigation period meet the requirements to kill all the pathogenic organisms.

TABLE 11.1: FORMALDEHYDE FUMIGATION CONCENTRATION

Purpose	Concentration of fumigation	Time required for fumigation (min)	Neutralizer
Hatching eggs immediately after laying	3X	20	No
Eggs in setter (1st day only)	2X	20	No
Chicks in hatchers	1X	3	Yes
Incubator room	2X	30	No
Hatcher between hatches	3X	30	No
Hatcher room chick room between hatches	3X	30	No
Wash room	3X	30	No
Chick boxes, Pads	3X	30	No
Truck	5X	20	Yes

(Source: Canadian Dept. Agr., Hatchery Sanitation, 1970)

DIRECTIONS FOR ITS USE:

1. Fumigation with 3X concentration of formaldehyde for 20 minutes will kill about 98.0% microorganisms on the egg shell. Higher percentage of organisms killed during summer season.
2. Formaldehyde gas is toxic to developing embryos particularly between 24 and 96 hrs of age when the chicks are pipping the

shell. To prevent the embryo from being weakened, eggs should be fumigated only as soon as they are placed in the incubators.

3. Formaldehyde fumigation colours the down of the chicks.
4. After the chicks have been removed from the hatchers and the trays washed, the hatchers, trays, hatcher room and washroom should be fumigated.

DRAWBACKS OF FORMALDEHYDE:

1. It is highly volatile, has penetrating odour and caustic action.
2. It is extremely irritant to the conjunctiva and mucus membranes.

EFFICACY AND USE OF DISINFECTANTS:

The efficacy of various disinfectants varies as the type organism (Table 11.2) and as per the area of hatchery to be disinfected (Table 11.3)

TABLE 11.2: PROPERTIES OF DISINFECTANTS

Property	Chlorine	Iodine	Phenol	QAC	Formaldehyde
Bactericidal	+	+	+	+	+
Bacteriostatic	-	-	+	+	+
Fungicidal	-	+	+	±	+
Virucidal	±	+	+	±	+
Toxicity	+	-	+	+	+
Action on organic matter	++++	++	+	+++	+

(Source: Canadian Dept. Agr., Hatchery Sanitation, 1970)

TABLE 11.3: USE OF DISINFECTANTS FOR VARIOUS AREAS OF HATCHERY

Property	Chlorine	Iodine	Phenol	QAC	Formaldehyde
Hatchery equipment	+	+	+	+	+
Water disinfection	+	+	-	+	-
Personnel	+	+	-	+	-
Egg washing	+	-	-	+	+
Floor	-	-	+	+	+
Footbaths	-	-	+	+	-
Rooms	±	+	±	+	+

(Source: Canadian Dept. Agr., Hatchery Sanitation, 1970)

CLEANING THE HATCHERY BETWEEN THE HATCHES:

Cleaning the hatchery between hatches is of primary importance. The process must be complete except for the setter and setter room, every equipment must be thoroughly vacuumed, scrubbed, disinfected and fumigated. Clean the setter room but do not fumigate it.

CLEANING THE HATCHERS:

1. Remove all racks, trays and clean them.
2. Vaccum the inside and outside of hatchers.
3. Wash the inside and outside of hatchers.
4. Scrub the inside wall with a suitable disinfectants.
5. Return all racks and clean trays from washroom to the hatcher and fumigate them with 3X concentration of formaldehyde.

CLEANING OF HATCHER ROOM AND CHICK ROOM:

1. Vaccum all debris from the floor and walls.
2. Wash the floors and walls and then disinfect them.
3. Wash and disinfect remaining equipment.
4. Fumigate rooms and equipment with 3X concentration of formaldehyde.

CLEANING THE HATCHER TRAYS:

All hatcher trays, carts and racks should be moved to the washroom and washed thoroughly dipped in a disinfecting solution moved to the clean room to dry and moved back to hatchers where they are fumigated with 3X concentration of formaldehyde.

CLEANING THE WASHROOM:

1. After all hatching trays and portable equipment have been washed and disinfected and taken out for fumigation; remove all debris from the washroom. Empty drain trap. Either incinerates the material removed or place it in plastic bags and remove from the hatchery.
2. Next wash and disinfect the ceiling, walls and floors.
3. Fumigate the room with formaldehyde with 3X concentration.

Steps to be taken for preventing the spread of infection:-

In order to reduce the possibility of disease as well as to reduce the spread of infection, the following steps should be adopted.

1. **Isolation and disease free chicks:** The hatchery should take every precaution to avoid bringing in infection. In this regard visitors should not be allowed. Showers and clothing of employees should be separate.
2. **Clean eggs:** Only clean eggs should be used for hatching and they should be collected and stored in clean equipment and containers.
3. **Sanitizing eggs:** All hatching eggs should be sanitized as soon as possible after collection and stored in clean equipment and containers.

Fumigation of eggs and incubators is an essential part of hatchery sanitation programme when improperly done, fumigation can be a hazard, and hence it should always be done by or under the supervision of an experienced person. Routine pre incubation fumigation of hatching eggs on the farm is highly recommended for eliminating Salmonella infection from the flocks. Each egg entering the hatchery should be subjected to pre incubation fumigation as soon as possible after its collection from the nest.

High levels of formaldehyde gas destroy Salmonella organisms on shell surfaces if used as soon as possible after the eggs are laid. Eggs for fumigation should be placed on racks so that the gas can reach the entire surface. Plastic trays used for washing market eggs are ideal for this purpose. A high level of formaldehyde gas is provided by mixing 1.2 cubic centimeters (cc) of formaline (37% formaldehyde) with 0.6 g of potassium permanganate ($KMnO4$) for each cuft of space in the cabinet. An earthen ware galvanized or enamel ware container having a capacity at least 10 times the volume of the total ingredients should be used for mixing the chemicals. The gas should be circulated within the enclosure for 20 minutes, and then expelled to the outside. Humidity for this method of pre incubation fumigation is not critical but the temperature should be maintained at approximately $70^{O}F$. Eggs should be set as soon as possible after fumigation and extra care taken to ensure that they are not exposed to new sources of contamination.

Eggs should be routinely refumigated after transfer to the hatchery to destroy organisms that may have been introduced as a result of handling. Recommendations for loaded incubator fumigation vary widely, depending upon the make of the machine. Therefore, the method concentration and duration of fumigation should be in accordance with the manufacturer's instructions. Empty hatchers should be thoroughly disinfected and fumigated prior to each transfer of eggs from setters.

4. **Separate rooms:** Hatcheries should be so designed that there are separate rooms for egg receiving, incubation, hatching, chick holding and waste disposal.
5. **Proper waste disposal:** Waste disposal may be a problem and a major source of infection in the hatchery unless it is properly handled. Incineration is an effective means of disposal of hatchery waste from the stand point of sanitation and disease control but it is an unprofitable method. The larger hatcheries

now process poultry waste into livestock feeds known as hatchery by product. It consists of infertile eggs, eggs with dead embryos and unsalable sexed male chicks. Hatchery by product is high in protein and calcium.

6. **Cleaning and disinfection:** In the hatcheries cleaning and disinfection should include the following steps:

   a. Remove trays and all controls and fans for separate cleaning. Thoroughly wet the ceiling, walls and floors with a stream of water, then scrub with a hard bristle brush. Rinse until no deposits are on the walls particularly near the fan opening.
   b. Replace cleaned fans and controls. Replace trays preferably while still wet from cleaning and bring the incubator upto the normal operating temperature.
   c. Fumigate the hatcher before inserting eggs.
   d. If eggs are hatching and incubating in the same machine, clean the entire machine after each hatch. Use a vaccum cleaner to remove chick down from the egg trays.

12

Hatchery Operation

The operation of a chick hatchery involves the production of the large numbers of quality chicks from the hatching eggs. The chicks produced must be produced economically. The following steps are involved in the production of quality chicks:

12.1. SECURING HATCHING EGGS:

As the hatching egg is the raw material of the hatchery, it is essential to ensure the steady supply of the right quality eggs in the numbers required to meet the demand for chicks of the various types sold. There will of course, be local requirements which will have to be discovered and catered for if the full success of the hatchery is not always made and the normal procedure is for a flexible arrangement under which the parties agree-provided no untoward circumstances arise to supply and accept respectively so many eggs of certain breeds or crosses per week over a period with a general index of the anticipated distribution of deliveries. The hatchery usually agrees to pay a certain price as long as the supplier continues to deliver eggs of a requisite size, weight and quality to feed his breeding flock properly and take all possible steps to minimize the risk of desire.

a) BREEDER FARMS OWNED BY HATCHERY:

The breeder flock may be owned by hatchery owner or the farms owned by the farmer who has entered into an agreement with the hatchery to produce hatching eggs on contract. In the latter case the hatchery actually owns the breeders, the farmer is paid a contracted amount to produce the eggs. This system is particularly applicable when the complete poultry enterprise is integrated.

b) HATCHERY SECURES EGGS FROM FARMERS:

In certain parts of world poultry farmers own both the breeders and the commercial poultry farm. Most of these farm owners are involved with some type of contract to produce eggs for a hatchery in predetermined quantities. Most of times, the hatchery is involved in financing the flock owners. In many instances the eggs may be shipped to a hatchery in another country.

c) HATCHING EGG PRODUCING COMPANIES:

Here eggs are produced by an egg operating company that has a large number of poultry farms under contract to produce hatching eggs for it. These companies in turn have an agreement to sell a given number of hatching eggs each week to other hatcheries that are involved with chick hatching.

12.2. DELIVERY OF HATCHING EGGS TO THE HATCHERY:

Disease free or MG—negative or MS—negative chicks are brought to the hatchery by truck, by rail or air route. The following rules must be rigidly observed:

1. All the trucks must be disinfected and fumigated with formaldehyde gas before eggs are placed in them.
2. Only eggs that are MG-negative or MS-negative may be placed in the truck.
3. Truck drivers and their helpers must shower and change in to clean clothes, head gear and footwear before entering any truck involved with the egg transport. The record of eggs received by the hatchery must include.

    1. Source of eggs.
    2. Date of eggs received.

3. Breed or line of chicken involved.
4. House, pen or flock numbers.

12.3. STORAGE OF HATCHING EGGS:

The storage conditions for hatching eggs are of great importance. It is necessary to ensure a steady temperature. The temperature depends mainly on the length of storage. The optimum egg storage temperatures are:

3 to 4 days of storage: 18-20^{0}C
5 to 9 days of storage: 15-16^{0}C

The relative humidity in the room should be 75-80%. If eggs have to be stored for longer than 4 days, a separate storage facility should be provided so as to maintain right temperature and humidity conditions to the eggs.

Thus prior to setting, eggs should be cooled by placing them in a room with a temperature of 65^{0}F (18.3^{0}C) and at a relative humidity of 75% which is the threshold of embryonic development and kept at this temperature until shortly before being placed in the incubators.

12.4. POSITION OF EGGS DURING HOLDING PERIOD:

During a holding period of less than 10 days, eggs should be placed small end down in the trays or on the flats when held longer than 10 days, hatchability will be improved if eggs are held small end up. When the hatching eggs are held for less than 1 week before being set, there seems to be no need for turning the eggs. However, in the case of certain poultry breeding and genetic farms, it may be necessary to hold eggs for rather long periods. Rotating eggs from side to side over a 90^0 angle will then improve hatchability.

12.5. GRADING HATCHING EGGS:

Some eggs are graded for quality and size in the hatchery. The decision depends on the type of eggs, breed and whether they are the product of an integrated operation.

A) **Broiler Type Hatching Eggs:** Eggs from which commercial broiler chicks are to be hatched may or may not be graded. The following situations are involved:

   a) When broiler-chick production is part of an integrated operation, there is little to be accomplished by egg grading. The process is expensive thus eliminated by the most integrators.
   b) When broiler chicks are sold to poultry farmers, the quality of chicks is of utmost importance. Small chicks will not be accepted by the customers, therefore, many of the hatching eggs must be graded and the small eggs must be discarded. This process is particularly applicable during the first few weeks of production of eggs from breeder hens which produce small sized egg during this period.

B) **Egg Type Hatching Eggs:** Hatching eggs that are to be used for the production of commercial laying type pullet chicks generally are graded for size and quality.

C) **Breeder Type Hatching Eggs:** Certain poultry farmers are involved with developing and merchandising breeder type chicks. Under these circumstances egg weight and quality are of the greatest importance.

If it is necessary to sort hatching eggs by size, the process should be completed after the eggs have cooled in the egg holding room. The usual procedure is to grade and tray eggs as close to setting time as possible. When there is a uniform daily labour supply in the hatchery this may not be practical and eggs may have to be trayed daily and ahead of setting time. If this is the case, the trayed eggs should be placed in covered cabinets. Do not allow hatching eggs to be located in free flowing air as in front of fans which would increase the rate of evaporation thus causing rapid drying of egg contents.

12.6. WARMING EGGS PRIOR TO INCUBATION:

It may be advisable to warm eggs prior to placing them in the incubators. Hatching eggs should not be removed from the cool holding room and placed directly in the setters. Rather they should be warmed to room temperature first but not any temperatures above 75^{0}F (23.9^{0}C) or at the point where embryonic development will be initiated. The warming process may take from 4 to 6 hours depending on the temperature of the egg holding room. Placing cold eggs in the setters usually reduces the temperature within the machine until the freshly set eggs reach the incubating temperature. This cool environment delays the hatching time of the newly set eggs and lowers the hatchability time of the eggs already in the incubator. However, some setters are equipped with an extra set of heaters that engage until the setter temperature is brought up to normal.

12.7. SETTING HATCHING EGGS:

The time the chick trucks are to leave the hatchery will determine the time that the eggs should be set. Chicks should be at the customer's farm about 12 hours after the hatch is pulled. Chicks will have to be removed from the hatcher at 9.00 to 10.00 P.M. in as much as chicks should be delivered early the following morning. Eggs should be set a time that will allow chicks to hatch and dry prior to 9.00 to 10.00 P.M. This will mean the average egg setting time will be about 5.00 to 6.00 P.M.

When eggs from different breeds or with different incubation periods are being set, they should be placed in the incubator so that all chicks hatch at the same time, e.g., eggs from leghorns should be set later than those from meat type breeders. The most important factors in artificial incubation are temperature, humidity, turnings and the exchange of O2 and CO2. An accurate record should be kept of these factors. The easiest way is to provide each setter with a chart on which the information is recorded each day at regular intervals.

- Temperature: It is the most critical single factor for successful hatching. The optimum temperature required for the development of embryo for first 18 days is 37.5 to 37.7^{0}C.
- Humidity: Humidity inside the incubator is of great importance for the normal development of chick embryos. During the embryonic development in the incubator moisture is lost from the contents through the shell due to higher incubator temperature. Hence a proper humidity is essential for achieving better hatchability. The optimum RH ranges from 50 to 60% in the setter i.e., wet bulb reading of 84 to 86^{0}F.
- Ventilation: Embryos use oxygen for their metabolism and give off carbon dioxide. The best hatchability is obtained with 1% oxygen and up to 0.5% CO2 levels inside the setter.
- Position of eggs: The ideal position of an egg during incubation in setters is with broad end up and narrow end down. Thus the embryo's head would occupy a position in the large end of the egg for proper hatching.
- Turning of eggs: Hatching eggs must be turned during incubation. It is most essential during the first week but becomes less important later. Turning should be done regularly at least every 3-4 hours interval in the setter.

12.8. CANDLING OF EGGS:

It is advisable to check the growth of embryos within the eggs by candling of eggs on 19th day of incubating. This helps in removing or discarding the infertile, the dead in germs together with cracked and rotten eggs thus creating a more favourable environment in the hatcher for the good eggs. Infertiles and rots tend to explode in the hatcher and will contaminate the other eggs resulting in poor hatch and lower chick quality.

12.9. TRANSFERRING THE EGGS TO HATCHER:

By the end of 18th day the eggs must be transferred from setter to the hatcher. The hatcher must be cleaned and disinfected thoroughly before loading the new batch of eggs. Make sure the entire inside of the hatcher, trollys and trays are absolutely dry. The transfer of eggs should be carried out as fast as possible but with smooth motions.

A correct and steady temperature and humidity level in the hatcher is necessary for proper hatching. The heat production of the embryo increase throughout the whole period of incubation but the increase is most rapid over the last two days. This will result in an egg temperature 2^{0}C higher than the ambient air of the incubator. For this reason the temperature of hatcher is reduced by ½ to 1^{0}C to avoid late embryonic mortality from 37.5^0-37.7^{0}C to 36.1 to 37.2^{0}C.

Humidity in the hatcher should be maintained to about 75% to help the chick break through the shell membrane. It will prevent the beak of the chick from sticking to the shell and stop the chick from drying out. Such Relative humidity is maintained by keeping the wet bulb reading of 31.11^{0}C (88^{0}F) to 32.5 (90.5^{0}F) for last 3 days of incubation. An adequate exhaust system connected to the hatcher should make sure that no harmful high levels of CO_2 are reached in the machine. A good constant supply of fresh air is necessary to maintain the right level of O_2. Normal levels of CO_2 for the hatcher are 0.6 to 0.8% by volume. Maintaining a low ventilation rate until hatch is to stimulate CO_2 increase thereby achieving adequate levels prior to hatch.

Position of eggs in hatcher: Most commercial incubators provide for keeping the eggs in a horizontal position during the last 2 days of incubation in the hatcher. Although they will hatch as well if kept upright with the broad end up, this method has not been practical become of the additional space needed by the chicks in the trays once they liberate themselves from the shells.

Turning eggs during the last 2 days of incubation has no value and may be injurious to the embryo. There is no evidence to show that changing the position of eggs at the time they are transferred to the hatchers is detrimental to hatchability provided the transfer is not made too early starting with the 17th day of incubation, the embryo begins to position itself for hatching and the process may take more than 24 hours.

12.10. FUMIGATION OF HATCHER:

Fumigation of chicks generally is not recommended, However, with omphalitis outbreaks in the hatchery it may be necessary to get the disease under control. Formaldehyde fumigation colours the down of chicks a deep orange, often noticeable by the customer. Chick fumigation should be considered an emergency measure only. About 40 ml liquid formalin and 20 gm of potassium permanganate in a bowl is placed it on floor of machine for 3 minutes.

12.11. PULLING THE HATCH:

The process of removing the chicks from the hatcher is often called as the pulling the hatch. It involves the following:

a) **Drying the chicks:** Dehydration is a stress and a problem with newly hatched chicks. Excessive drying in the hatcher should be avoided. Chicks should be removed from the hatcher as soon as all are hatched and about 90% are dry. Further drying and hardening should be confined to chick boxes.
b) **Box filling with chicks:** When the chicks are removed from the hatcher and boxed, they should be moved to chick holding room where the temperature should be between 22 to 28°C preventing the chicks from cooling too rapidly.

12.12. HANDLING OF CHICKS:

The chicks should be stored in a well-ventilated area with the temperature kept around 22 to 28 °C and R.H. of 55%. Storage prior to dispatch should be kept as short as possible.

12.13. GRADING THE CHICKS:

Chick quality depends on a number of aspects influencing the egg from the time before it is laid until chick delivery at the rearing house. It starts with breeder nutrition, processing and transporting chicks. Only when every step in production cycle is achieved under the best possible conditions, highest chick quality standards will be achieved.

A definite standard of chick quality must be outlined. No chick below the minimal standard weight must be allowed to go to a customer. The standard should be the same for all breeds and functional in all seasons of the year. Do not cut quality when the hatch is poor.

Some standards for quality are:-

1. No of chick deformed.
2. No of unhealed navals.
3. Not dehydrated.
4. Down colour representative of the breed.
5. Stand up well, are lively.
6. Above a minimum weight.

However, the criteria for selection of the chicks are not easy to define. Chicks with abnormalities on the head, beak or legs are easy to recognize and are removal right away.

The most common reasons for down grading chicks are:-

Deformities: Spraddle legs, malformed beak, wary neck, open brain, missing eyes etc. Most of these originate from the faulty nutrition and too high incubation temperatures.

Unhealed naval: In certain chicks thread like structures hang out of naval which may be caused by too high humidity, older breeder flock and hatchery borne infections.

Yolk sac infection: Soft belly, large chicks with discoloured yolk sac and chicks show a distinct smell. It is caused by bacterial contamination in the eggs after laying and before hatching, wrong egg room condition, dirty nests, egg flats, setter trays etc.

Dehydration: Small, skinny chicks with pale colour of feet. Chicks have hatched early & spent too long time in the hatcher after hatching chicks die between 2 and 7 days.

Poor quality feathers: This is due to nutritional deficiency in breeder flock.

12.14. SEXING THE CHICKS:

The method of sexing depends on the breed. leghorns are vent or feather sexed, most brown egg varieties are colour sexed and broilers are colour sexed or feather sexed.

12.15. RECORDING THE DATA:

The following data should be recorded for each group of hatching eggs set.

1. Breed
2. Number of eggs set.
3. Number of quality chicks hatched.
4. Percentage of total hatchability.
5. Number of culls.
6. Percentage of culls.

12.16. PACKING CHICKS IN CHICK BOXES:

Chick boxes made of corrugated sheet are used for packing the day old chicks. Each box is subdivided into 4 chambers by corrugated dividers. This is to prevent the chicks from accumulating in one corner; some material to which they may clamp their toes should cover the bottom of the box. These chick boxes vary in size as follows:

Standard winter	22x18x6 inches	56x46x15 cm
Standard summer	22x18x6 inches	56x46x15 cm
Large size	24x20x6 inches	61x51x15 cm

The number of chicks placed in each box depends on the environmental temperature and the distance that the chicks have to reach. During winter season when the ambient temperature is below 21°C each box hold 100 chicks and in summer when the ambient temperature is above 21°C it holds 80 chicks each.

If the bamboo baskets are used a diameter of 48cms with a height of 13cms can accommodate 50 chicks comfortably. Chick boxes/baskets should not be covered with tarpaulins.

12.17. HARDENING THE CHICKS:

When chicks are first placed in the chick boxes they are soft in the abdomen, are not completely fluffed out and do not stand well. They must be hardened by leaving them in the boxes for 4 to 5 hours. Such hardening makes it easier to grade the chicks for quality and the chicks are easily vent sexed.

12.18. EXTRA CHICKS:

The hatchery owner added about 2% chicks to replace any dead chicks prior to the arrival of the chicks at the farm. This way any complications with regards to adjustments after they have been delivered are avoided.

12.19. CHICK DELIVERY:

Safe and sanitary delivery of the day old chicks is the last phase of hatchery operation. Throughout the world, most deliveries are by trucks, although other means of transportation such as rail and air are also used for the purpose. In some instances the customers may pick up his chicks at the hatchery, using his own means of transportation.

a) DELIVERING CHICKS BY TRUCKS:

Chicks should reach the customer's farm early in the morning. The weather will be cooler during this part of the day and also it would allow a full day for close observation of chicks by the owner. The size of the compartments of truck in which the chick boxes are placed will vary. Some trucks hold 10,000 to 50,000 chicks. The size of the truck will be determined by the size of the hatchery.

b) SHIPPING CHICKS BY AIR:

Many chicks are shipped long distances by air. Special recommendations and instructions are necessary as follows:

- No smaller than 18 x 24 x 7 inch boxes should be used having a volume of 2800 $inch^3$ which is important for air shipment.
- Punch out holes in the boxes.
- 80 chicks should be packed in a standard box when the temperature is above 21°C and 100 chicks when the temperature is below 21 ^{O}C.
- Bring chicks well in time to airport.
- If airline transfers are to be made enroute, do not schedule more than 6 hours for the transfer.
- Book direct flight if possible.

INSTRUCTIONS AT THE AIRPORT:

Probably more chicks die or are damaged at the airport than in the plane. Extreme precautions must be taken as follows:

- The airport authorities must be informed about the chicks on board.
- Keep chicks in the shade.
- Keep chicks away from drafts.
- Do not allow chick boxes to stand outside in hot summer or cold weather.
- Do not cover chick boxes with a tarpaulin.
- Never place boxes in a corner of a room.
- Keep in a well-ventilated room inside the cargo building.
- Don't stack boxes over eight boxes.
- Never allow the boxes to become wet. They will collapse.
- Don't stack other cargo on top of chick boxes leave air space around the boxes.
- Keep boxes level at all times.

Under optimal conditions chicks can withstand the transportation stress over 48 hours without any significant mortality because of energy obtained from yolk sac. The two key factors that will have a negative influence on chick quality during transportation are overheating and dehydration.

13

Project Report for Establishment of A Broiler and Layer Hatchery

13.1. PROJECT REPORT FOR ESTABLISHMENT OF A BROILER HATCHERY

The broiler meat consumption in India is going up by about 10% every year. To meet this demand, sufficient number of day old broiler chicks are needed for which more hatchers are essential. Hence this project is proposed in a broiler deficit area by the promoters.

OBJECTIVES:

c) To start a broiler hatchery producing about 32,700 broiler chicks per week.

d) To maintain a broiler breeding stock in 1+2 batches with 5000 females parents per batch; with 26 weeks batch interval.

TECHNICAL DETAILS OF THE PROJECT

Type of parent stock	: Broiler strain
No. of batches	: 1 batch grower and 2 batches breeders at any time (1+2 system).
Batch size	: 5500 female and 600 male parent stock will be purchased per batch.
Housing system	: Growers on deep litter and breeders in cages.

Batch interval	: 26 weeks i.e., 2 batches per year.
Culling age	: 70-75 weeks.
Mating system	: Artificial insemination.
Cost of day old parent chick	: Rs. 100/-
Feed	= Own feed @ Rs. 7/kg.
Grower floor space	= 2 sqft. / Bird
Breeder hen cage space	= 135 sq. inches/hen.
	= 15" depth x 18"front cage for 2 hens.
Breeder house space (Including cage)	:2 sqft./Bird
Feed consumption/bird Up to 24 weeks of age	: 12 kg/Bird
Feed consumption during laying period (25-70 weeks of age)	= 58 kg (180kg/Bird/day)
Mortality and culling during growing period (0-24 weeks	= 10%
Mortality and culling during breeding period (25-70 weeks)	= 12%
Total eggs per hen housed	= 190
Settable (hatching eggs per hen housed)	= 180
Saleable chicks/hen housed	= 150
Artificial insemination will be done using freshly harvested semen once in 5 days (Average weekly hatching eggs produced	= 38500
Average weekly production of saleable broiler chicks	= 85% above = 32700
Average weekly saleable table eggs (10 eggs per hen in 46 weeks)	= 2,000
Culled birds (Hen + Cock) for sale after 10 + 12% mortality	= 4830/batchx2batch/year
	= 9660
Mannure available for sale @ 50 kg per hen/cock housed	= 550 tonnes.
Empty feed bags for sale at 70 kg feed/ bird with one saleable bag for 100 kg feed	= 0.7 bags x 11000 birds
	= 7700 bags

BANK LOAN REQUIRED:

The total cost of the project is Rs. 116.36 lakhs of which the share of owner will be land cost + Rs. 29,10,000/ and the bank finance required will be Rs. 87.28 lakhs. This amount along with interest will be repaid in 5 years period after one year holiday period.

(I) Capital Investment

(A) Lands and Allied Expenses	**Cost (Rs. in lakhs)**
(1) Land development charges including fencing, borewell, overhead tank, pipeline, farm roads, watch man booth etc.	2.00
(B) Buildings:	
(1) Brooder cum grower deep litter house for 6000 bird @ 2 sqft / bird = 400′ x 38′ size @ Rs. 80/ sqft	9.60
2) Cage breeder houses (2 Nos.) each to house 5000 hens and 500 cocks with 2 sqft/bird = 22000 sqft 30 x 739 @ Rs. 80/ sqft	17.60
3) Feed mill 2000 sqft + store rooms (2000 sqft) @ Rs. 150 sqft	6.00
4) Hatchery building (2400 sqft) @ Rs. 200/sqft	4.80
5) Worker's quarters @ 200 sqft for 6 person @ Rs. 200/ sqft	2.40
6) Manager's quarter 800 sqft with two supervisor's quarters @ 500 x 2 = 1000 sqft @ Rs. 400/sqft	7.2
7) Office (400 sqft) @ Rs. 400 sqft.	1.6
Total Expenditure	**Rs. 51.20**

Details of Expenditure	**Cost (Rs. in lakhs)**
(C) Cost of Equipment	
1) Cost of broading & grower feeders & drinkers For 6000 birds @ Rs. 15/bird	0.90

2) Cost of breeder hen cages (1000 hens) + 1000 cocks @ Rs. 100/bird	11.00
3) Feed mill with electrical accessories	4.00
4) Cost of 4 setters of 30,000 capacity each @ Rs. 2 lakh each.	8.00
5) Cost of 2 hatchers each of 10,000 capacity @ Rs. 1 lakh each	2.00
6) Cost of one 50 KV and one 25 KV generators with room and accessories.	5.00
7) Cost of one tonne chick delivery van	5.00
8) Cost of office @ other farm equipment.	1.00
Total Cost of Equipment	**36.90**

D) Working Capital

Details of Expenditure	**Cost (Rs. in lakhs)**
1. Cost of 6100 parent stock/Batch x 3 batches @ Rs. 85 each	15.56
2. Cost of feed for 2 batches up to 24 weeks (Average 5800 birds x 2 batches x 12 kg feed/Bird x Rs. 7/kg)	9.74
3. Miscellaneous cost of 2 batches for 6 months @ Rs.3/bird/month = 11,000 x 18	1.98
4. Office and hatchery expenditure	1.00
Total Working Capital	**28.28**

Total Project Cost and Bank Finance Needed (Rs. in lakhs)

Details of Expenditure	**Total cost**	**Share of the promoter**	**Bank finance needed**
Cost of buildings	51.20	12.80	38.40
Cost of equipment	36.90	9.23	27.67
Working capital	28.28	7.07	21.21
Total	**116.38**	**29.10**	**87.28**

The bank loan will be repaid in 5 years period along with interest after one year holiday period as follows:

I	Year	=	Nil	+	Interest only
II	Year	=	7.28	+	Interest
III	Year	=	20.00	+	Interest
IV	Year	=	20.00	+	Interest
V	Year	=	20.00	+	Interest
VII	Year	=	20.00	+	Interest
	Total	=	87.28		
	Rate of Interest = 15%				

Annual Recurring Expenditure (Rs. in lakhs)

Year	Chick cost Rs. 85 each	Feed cost (2.3 tonnes/ day	Misc. farm expenses (Rs.3/b/m)	Hatchery expenses (Rs.0.25/ egg)	Office & marketing expenses (Rs.0.25/ chick)	Total expenditure
1.	(10.37)	(9.74)	(1.98)	(0.50)	(0.50)	(23.09)
	-	12.71	0.99	1.08	0.85	15.63
2.	(5.19)	-	-	-	-	(5.19)
	(5.18)	58.77	5.94	5.00	4.24	79.13
3.	10.37	58.77	5.94	5.00	4.24	84.32
4.	10.37	58.77	5.94	5.00	4.24	84.32
5.	10.37	58.77	5.94	5.00	4.24	84.32
6.	10.37	58.77	5.94	5.00	4.24	84.32
7.	10.37	58.77	5.94	5.00	4.24	84.32

Values in parenthesis go towards working capital. Hence it does not come under annual recurring expenditure for calculating profits and losses.

Annual Gross Returns, Net Profits (Rs. in Lakhs) and Benefit: Cost Ratio

Details	Year						
	1	2	3	4	5	6	7
1. By sale of chicks at Rs. 10/ chick	40.00	170.04	170.04	170.04	170.04	170.04	170.04
2. By sale of culled birds @ Rs. 50/—each	-	4.83	4.83	4.83	4.83	4.83	4.83
3. By sale of eggs unfit for hatching @ Rs. 1 each	0.30	1.04	1.04	1.04	1.04	1.04	1.04
4. By the sale of manure at Rs. 200/tonne	0.28	1.10	1.10	1.10	1.10	1.10	1.10
5. By sale of empty feed bags at Rs. 7/bag	0.14	0.54	0.54	0.54	0.54	0.54	0.54
6. Total Gross Receipts	**40.72**	**177.55**	**177.55**	**177.55**	**177.55**	**177.55**	**177.55**
7. Total gross expenditure	15.63	79.13	84.32	84.32	84.32	84.32	84.32
8. Receipts over expenditure	25.09	98.42	93.23	93.23	93.23	93.23	93.23
9. Depreciation @ 5% on buildings and 10% on equipment	-	6.25	6.25	6.25	6.25	6.25	6.25
10. bank loan and interest repayment	13.09	20.37	32.00	29.00	26.00	23.00	-
11. Net Profits	**12.00**	**71.80**	**54.98**	**57.98**	**60.98**	**63.98**	**86.98**
12. Benefit: Cost ratio	1.42	1.68	1.45	1.49	1.52	1.56	1.96

Bank Loan Repayment Schedule (Rs. in lakhs)

Year	O.B. of loan	Interest	Total	Repayment	Closing balance
1.	87.28	13.09	100.37	13.09	87.28
2.	87.28	13.09	100.37	20.37	80.00
3.	80.00	12.00	92.00	32.00	60.00
4.	60.00	9.00	69.00	29.00	40.00
5.	40.00	6.00	46.00	26.00	20.00
6.	20.00	3.00	23.00	23.00	Nil
7.	Nil	-	-	-	-
Balance sheet	**87.28**	**56.18**	**143.46**	**143.46**	**Nil**

Cash Flow Statement and Internal Rate of Return (Rs. in lakhs)

Details of cash transaction	At start	At the end of the year						
Land cost	5.00	-	-	-	-	-	-	-
Promoter's fund	29.10	-	-	-	-	-	-	-
Borrowings	-	87.28	-	-	-	-	-	-
Cash inflow	-	40.72	177.55	177.55	177.5	177.5	177.5	177.5
Cash outflow	-	15.63	79.13	84.32	84.32	84.32	84.32	84.32
Repayment of loan	-	13.09	20.37	32.00	29.00	26.00	23.00	-
Depreciation	-	-	6.25	6.25	6.25	6.25	6.25	6.25
Net profit	-	12.00	71.80	54.98	57.98	60.98	63.98	86.98
Net assets cumulative	34.10	46.1	117.9	172.88	230.86	291.84	355.82	442.8
Benefit cost ratio	-	1.42	1.68	1.45	1.49	1.52	1.56	1.96
Interest rate of return (%)	-	10.3	67.95	64.43	67	69.58	72.16	74.74

13.2. PROJECT REPORT FOR ESTABLISHMENT OF A LAYER HATCHERY:

1. Objectives:

a) To start a layer hatchery capable of producing 30,000 day old egg-type pullet (female) chicks per week.
b) To maintain 1 + 3 batches; with a batch interval of 20 weeks i.e., 2.6 batches per year.
c) To go for forward integration, by maintaining own layer farms and marketing of table eggs at a later date.

2. Technical Details of The Project:

Type of parent stock	: Layer strain
Housing system	: Deep litter
No of batches in the farm	: 1 batch of grower + 3 batches of breeder at any time.

Batch size	: 5500 female + 600 male parent stock.
Batch size at start of lay	: 5000 + 500
Batch size at culling time	: 4500 + 450
Batch interval	: 20 weeks i.e., 2.6 batches per year.
Culling age	: 76 weeks of age
Hatching eggs collection period	: 25 to 76 weeks of age = 52 week period
Mating system	: Natural mating with 1:10 sex ratio
Feed	: Own feed
Grower floor space	: 1.5 sq.ft./bird
Breeder floor space	: 2.6 sq.ft./bird
Feed consumption/bird during growing period	:7 kg
Feed consumption during laying period	: 110g/bird/day
Average daily feed consumption for the full 1 + 3 batches = 2000 kg	
Mortality during growing period	:10%
Mortality during laying period	: 10%
Total eggs / hen housed	: (20-76 weeks) = 290 eggs
Hatching eggs/hen housed (25-76 weeks) = 270 eggs	
Percent total hatchability	: 90%
Saleable pullet (female) chicks/hen housed (25 to 76 weeks) = 115	
Average weekly production of hatching eggs	: 70,000
Average weekly production of table eggs (unfit for hatching) = 5,000	
Average weekly production of saleable pullet chicks	: 30,000
Average culled birds for sale/year: 12870= (4950/batch x 2.6 batch)	
Average manure produced/year @ 30kg/bird housed x 14300 birds: 429 tonnes	
Empty feed bags for sale: 20 bags/days x 365 days = 7,300 bags	
Male chick for sale: 20 bags/day: 30,000/week	
Selling price of culled bird	: Rs. 25 each

Feed cost	: Rs.7/kg
Day old parent chick cost	: Rs. 150 each
Day old female chick cost	: Rs. 15/chick
Day old male chick cost	: Rs. 0.25/chick
Cost of manure	: Rs. 200/tonne
Cost of empty feed bags @ Rs.7/bag	
Cost of table eggs (eggs unfit for hatching)	: Rs. 0.75/egg
Miscellaneous cost of bird	: Rs.3/bird/month
Hatching cost	: Re. 0.25/egg set
Marketing and office cost Rs. 0.50/female chick sold	

Bank Loan Required and Repayment Schedule:

The total cost of the project is Rs. 150.24 lakhs of which the promoter's share i.e., margin money will be land cost Rs. 37.56 lakhs and the bank finance required will be Rs. 112.68 lakhs. This amount along with 15% interest will be repaid in 5 year period after one year holiday period.

Cost of Buildings and Other Civil Works	(Rs. in lakhs)
1. Hand development charges in including farm roads well, fencing, over head water tank etc.	3.00
2. Brooder cum grower house for 6100 chicks (5500 female + 600 male parent stock) at 1.5 sq.ft./bird @ 9150 sq.ft. at Rs. 80/sq.ft.	7.32
3. Brooder house after 3 batches: each for 5000 + 500 birds x 2 sq.ft./bird = 33000 sq.ft. @ Rs.80/sq.ft.	26.40
4. Feed mill 2000 sq.ft. + misc. stock rooms = 2000 sq.ft. for different products @ Rs. 150/sq.ft.	6.00
5. Hatchery building 3600 sq. ft. @ Rs. 150/sq.ft.	5.40
6. Quarters for workers (8 persons) x 200 sq.ft. each x Rs. 200/sq.ft.	3.20
7. Supervisor's (500 sq.ft. x 3) & manager's (800 x 1) quarters 4 Nos = 2300 sq.ft. x Rs. 400/sq.ft.	9.20
Total Expenditure on Civil Works	**60.52**

1. Cost of brooding and growing equipment for 6000 birds at Rs.15/bird.	0.90
2. Cost of breeder equipment for 5500 x 3 batches x Rs.20/bird.	3.30
3. Feed mill with electrical accessories.	4.00
4. Cost of 8 setters each of 30,000 egg capacity @ Rs.2 lakhs each.	16.00
5. Cost of 4 hatchers each of 10,000 egg capacity @ Rs. 1 lakhs each.	4.00
6. Cost of one 60 KV and one 35 KV generators.	6.00
7. Cost of one 4 tonne chick delivery van.	5.60
8. Cost of office and other miscellaneous equipments.	1.00
Total equipment cost	**40.20**

Working Capital	**(Rs. in lakhs)**
1. Cost of 6100 chicks x 3 batches of parent stock @ Rs. 125/each.	22.88
2. Feed cost for 3 batches for different periods i.e., 1 year, 32 works and 12 weeks period respectively = 179.5+91.5+24 = 295 tonnes @ Rs. 7000/tonne.	20.65
3. Miscellaneous cost of 3 batches for 3 different periods @ Rs.3/bird/month = 205.5+139.5+54	3.99
4. Office & hatchery expenses for Ist year.	2.60
Total Working Capital	**49.52**

Total Project Cost, Margin Money and Bank Finance Needed (Rs. in Lakhs)

	Item of expenditure	Total cost	Share of promoter	Bank finance needed
1.	Cost of building	60.52	15.13	45.39
2.	Cost of equipment	40.20	10.05	30.15
3.	Working capital	49.52	12.38	37.14
	Total	**150.24**	**37.56**	**112.68**

The bank loan will be repaid in 5 years period along with 15% interest after one year holiday period as follows:

I	Year	=	Nil	+	Interest only
II	Year	=	12.68	+	Interest
III	Year	=	25.00	+	Interest
IV	Year	=	25.00	+	Interest
V	Year	=	25.00	+	Interest
VI	Year	=	25.00	+	Interest
	Total	=	**112.68**	+	**Interest**

Year	Chick cost (Rs. 125 each x 2/3 batches) per annum	Feed cost (2 tonnes/day) x Rs. 7000/ tonne	Misc. farm expenses (Rs.3/bird/ month)	Hatching expenses (Rs.0.25/ eggs set)	Office & marketing expenses (Rs.0.50/ chick)	Total expenditure
1	(22.88)	(20.65)	(3.99)	(21.00)	(1.00)	(49.52)
2	15.25	49.42	6.66	8.34	7.15	86.82
3	22.88	51.10	7.20	9.10	7.80	98.08
4	22.88	51.10	7.20	9.10	7.80	98.08
5	15.25	51.10	7.20	9.10	7.80	90.45
6	22.88	51.10	7.20	9.10	7.80	98.08
7	22.88	51.10	7.20	9.10	7.80	98.08

Values in parenthesis goes towards working capital, hence there is no annual recurring expenditure during first year.

Annual gross returns, net profits (Rs. in lakhs) and benefit: cost ratio

		year						
Sl. No.	**Details**	**1**	**2**	**3**	**4**	**5**	**6**	**7**
1	By sale of pullet chicks @ Rs. 15 each	36	207	234	234	234	234	234
2	By sale of culled birds @ Rs.25 each with 2.6 batches/ year	-	3.22	3.22	3.22	3.22	3.22	3.22
3	By sale of male chicks @ Re.0.25 each	0.6	3.45	3.9	3.9	3.9	3.9	3.9
4	By sale of eggs for table purpose @ Re. 0.75 each	0.3	1.73	1.95	1.95	1.95	1.95	1.95
5	By sale of manure at Rs. 200/tonne	0.05	0.71	0.8	0.8	0.8	0.8	0.8
6	By sale of empty feed bags @ Rs. 7/bag	0.21	0.45	0.51	0.51	0.51	0.51	0.51
7	Total gross receipts	37.16	216.56	244.38	244.38	244.38	244.38	244.38
8	Total gross expenditure	-	86.82	98.08	98.08	90.45	98.08	98.08
9	Receipts over expenditure	37.16	129.74	146.3	146.3	153.93	146.3	146.3
10	Depreciation on buildings @ 5%	-	7.05	7.05	7.05	7.05	7.05	7.05
11	Bank loan & interest repayment	16.9	29.58	40	36.25	32.5	28.75	-
	Net profit	**20.26**	**93.11**	**99.25**	**103**	**114.38**	**110.5**	**139.25**

Bank loan repayment schedule (Rs. in lakhs)

Year	O.B. of loan	Interest 15%	Total	Repayment	Closing balance
1	112.68	16.9	129.58	16.9	112.68
2	112.68	16.9	129.58	29.58	100
3	100	15	115	40	75
4	75	11.25	86.25	36.25	50
5	50	7.5	57.5	32.5	25
6	25	3.75	28.75	28.75	Nil
7	Nil	-	-	-	-
Balance sheet	**112.68**	**71.3**	**183.98**	**183.98**	**Nil**

Cash flow statement, benefit: cost ratio and internal rate of return (Rs. in lakhs)

Details of cash transaction	At start	At the end of the year						
		1	2	3	4	5	6	7
1. Hand cost	5	-	-	-	-	-	-	-
2. Promoter's share	37.5	-	-	-	-	-	-	-
3. Borrowings	-	112.68	-	-	-	-	-	-
4. cash in flow	-	37.16	216.56	244.38	244.38	244.38	244.38	244.38
5. Cash outflow	-	-	86.82	98.08	98.08	90.45	98.08	98.08
6. Repayment of loan	-	16.9	29.58	40	36.25	32.5	28.75	-
7. Depreciation	-	-	7.05	7.05	7.05	7.05	7.05	7.05
8. Net profits	-	20.26	93.11	99.25	103	114.38	110.5	139.25
9. Net assests-cumulative	42.56	62.82	155.93	255.18	358.18	472.56	583.06	722.31
10. Benefits: cost ratio	-	2.2	1.76	1.68	1.73	1.88	1.83	2.32
11. Internal rate of return (%)	-	120	80.27	77.54	76.87	93	87	142.1

14

Annexures

Annexure I: Dry and wet bulb readings corresponding to varying relative humidities

Parameter	Dry bulb		Wet bulb reading					
			60 percent relative humidity		70 percent relative humidity		80 percent relative humidity	
	°F	°C	°F	°C	°F	°C	°F	°C
Storage range	50	10	44.2	6.8	45.8	7.4	47.2	8.2
	52	11.1	46	7.8	47.6	8.3	49.1	9.3
	54	12.2	47.7	8.4	49.4	9.4	51	10.5
	56	13.3	49.5	9.6	51.2	10.6	52.9	11.4
	58	14.3	51.3	10.7	53	11.5	54.7	12.1
	60	15.3	53	11.5	54.8	12.2	56.6	13.3
Incubator range	99	37	86.6	30.3	90.2	32.3	93.2	34
	99.5	37.2	87.2	30.6	90.7	32.4	93.6	34.2
	100	37.5	87.7	30.9	91.2	32.8	94.1	34.6
	101	38.1	88.5	31.2	92.1	33.2	95.1	35
	102	38.7	89.3	31.8	93	33.9	96	35.4

Annexure II: Incubation period of eggs from different chicken species

Type of bird	Age (days)
Bob white	23
Chicken	21
Canary	14
Ducks	28
Geese	31
Guinea fowl	27
Japanese quail	18
Pheasant	28
Pea fowl	28
Parrot	25
Swan	42
Partridge	23
Pea cock	28
Turkey	28
Muscovy	35
Ostrich	42
Emu	55

Annexure III: Specifications for a standard egg

Weight (g)	56.7
Volume (cm^3)	63.0
Specific gravity	1.09
Long circumference (cm)	15.7
Short circumference (cm)	13.7
Shape index	74.0
Surface area (cm^2)	68.0

Annexure IV: Vaccination schedule for the layer and broiler parents

Age	Disease	Vaccine	Dose/ Mode of administration	Remarks
1 day	Marek's	HVT strain	0.2 ml S/C	At the hatchery
7-10 days	Ranikhet disease	F1 strain	1-2 drops I/O or I/M	At farm level
14 days or 18-21 days	Gumboro	Georgia strain	Drinking water	Do
25-30 days	Ranikhet disease	Lasota strain	Drinking water	Do
5 weeks	Gumboro	Intermediate strain	Drinking water	Do
6-7 weeks	Fowl pox	Fowl pox vaccine	Rubbing in wing webs	Do
8 weeks	Ranikhet disease	R2B vaccine	Intramuscular	Do
12 weeks	Infectious bronchites	IB vaccine	Intra nasal	Do
14 weeks	Fowl pox	Fowl pox vaccine	Rubbing in wing webs	Do
16 weeks	Ranikhet disease	R2B strain	Intramuscular	Repeated for commercial layers
18 weeks	Ranikhet disease	Killed vaccine	Intramuscular	For layer parents only
19 weeks	Gumboro	Killed vaccine	Intramuscular	For layer parents only
22 weeks	Ranikhet disease	Killed vaccine	Intramuscular	For broiler parents only
23 weeks	Gumboro	Killed vaccine	Intramuscular	For broiler parents only
45-50 weeks	IBD & RD	Killed vaccine	Intramuscular	For both parents

Annexure V: Suggested nutrient levels for broiler breeder flocks

Nutrient	Starter	Grower	Pre Breeder (20-24 weeks)	Breeder
Protein %	20-21	16-17	17-18	18-19
ME Kcal/kg	2800	2650	2700	2700
Fat%	3	3	3	3
Calcium%	1.2	1.2	2.2	3.25
Phosphorus available %	0.45	0.45	0.45	0.45
Fibre%	5	3-6	7	7
Linoleic acid%	1.3	1.3	1.4	1.4
Lysine%	1.0	0.70	0.75	0.80
Methionine%	0.45	0.30	0.33	0.35
Methionine+cysteine%	0.80	0.58	0.62	0.65
Sodium%	0.15-0.2	0.15-0.2	0.15-0.2	0.15-0.2
Chloride%	0.15-0.2	0.15-0.2	0.15-0.2	0.15-0.2
Potassium%	0.40-0.50	0.40-0.50	0.40-0.50	0.40-0.50
Iron, mg/kg	80	80	80	80
Copper, mg/kg	20	20	20	20
Manganese, mg/kg	80	80	100	100
Zinc, mg/kg	60	60	80	80
Selenium, mg/kg	0.2	0.2	0.2	0.2
Vitamin A, IU/kg	12,000	12,000	15,000	15,000
Vitamin D3, IU/kg	1,500	1,500	3,000	3,000
Vitamin E, IU/kg	20	20	30	30
Vitamin K, IU/kg	2	2	2	2
Thiamin, mg/kg	4	4	4	4
Riboflavin, mg/kg	6	6	10	10
Pantathonic acid, mg/kg	15	15	20	20
Niacin, mg/kg	50	50	60	60
Pyridoxine, mg/kg	8	8	8	8
Biotin, mg/kg	0.15	0.15	0.20	0.20
Choline, mg/kg	1500	1500	1500	1500
Folic acid, mg/kg	2	2	2	2
Vitamin B12, mg/kg	15	15	20	20

Annexure VI: Nutrient specifications for egg type breeders on as fed basis (88% of dry matter)

Nutrient Age (weeks)	Chicks	Growers	Growers	Prebreeders	Female breeders	Male
	0-6 wk	7-12 wk	13-18 wk	19 wk		From 19 wk
Feed intake g/b/d					105	80
Moisture, %	12	12	12	12	12	12
ME, Kcal/kg	2850	2850	2850	2850	2700	2850
Protein %	20	16.6	14.4	16.6	17	15
Arginine %	1.16	0.85	0.74	0.85	0.94	0.77
Histidine %	0.33	0.24	0.21	0.24	0.22	0.22
Isoleucine %	0.76	0.57	0.49	0.57	0.74	0.51
Leucine %	1.26	0.93	0.80	0.93	1.01	0.83
Lysine %	1.05	0.78	0.67	0.78	0.85	0.70
Methionine %	0.49	0.37	0.32	0.37	0.40	0.33
Methionine + cystine %	0.89	0.67	0.58	0.67	0.71	0.60
Phenyl alanine %	0.68	0.50	0.42	0.50	0.71	0.44
Phenyl alanine + tyrosine %	1.24	0.92	0.79	0.92	1.02	0.82
Threonine %	0.69	0.52	0.45	0.52	0.57	0.47
Tryptophan %	0.19	0.14	0.12	0.14	0.16	0.13
Valine %	0.84	0.62	0.53	0.62	0.86	0.55
Linoleic acid %	1.00	1.00	1.00	1.00	1.00	1.00
Calcium %	0.90	0.80	0.70	2.00	3.50	0.60
Chloride	0.18	0.15	0.14	0.15	0.17	0.14
NPP %	0.45	0.40	0.40	0.40	0.27	0.30
Sodium	0.18	0.15	0.15	0.15	0.16	0.15

Annexure VII: Nutrient specifications for broiler breeders on as fed basis (88% dry matter)

Nutrient Age (weeks)	Chicks	Growers	Growers	Prebreeders	Female breeders	Male
	0-6 wk	7-12 wk	13-18 wk	19 wk		From 19 wk
Moisture, %	12	12	12	12	12	12
ME, Kcal/kg	2850	2850	2850	2850	2700	2850
Protein %	19	16	14	16	16	14
Arginine %	1.11	0.89	0.78	0.89	0.78	0.78
Histidine %	0.30	0.24	0.21	0.24	0.21	0.20
Isoleucine %	0.73	0.58	0.51	0.58	0.61	0.48
Leucine %	1.29	1.03	0.90	0.96	0.85	0.78
Lysine %	1.00	0.80	0.70	0.80	0.70	0.65
Methionine %	0.46	0.37	0.32	0.37	0.32	0.32
Methionine + cystine %	0.84	0.67	0.59	0.67	0.60	0.56
Phenyl alanine %	0.65	0.51	0.45	0.51	0.50	0.41
Phenyl alanine + tyrosine %	1.19	0.95	0.83	0.95	0.83	0.77
Threonine %	0.67	0.53	0.47	0.54	0.47	0.44
Tryptophan %	0.19	0.15	0.13	0.15	0.13	0.13
Valine %	0.78	0.62	0.55	0.63	0.65	0.57
Linoleic acid %	1	1	1	1	1	1
Calcium %	0.90	0.80	0.70	1.20	3.00	0.60
Chloride, %	0.18	0.15	0.14	0.15	0.17	0.14
Phosphorus, non phytin %	0.45	0.40	0.40	0.40	0.27	0.30
Sodium	0.18	0.15	0.15	0.15	0.16	0.15